The secret military remote perception skill anyone can learn

Paul H. Smith, PhD

THE ESSENTIAL GUIDE TO REMOTE VIEWING

The secret military remote perception skill anyone can learn

Intentional Press
Box 2636, Cedar City, Utah 84720
editor@intentionalpress.com
intentionalpress.com
(866) 229-7847

Ordering Information:
Quantity sales. Special discounts are available on quantity purchases by corporations, individuals, associations, and others. For details, contact the publisher at the address above.

Second Printing, December, 2017

ISBN 9781938815010

Cover Artwork: Kris Mangrum
Cover Design: Josh Sisco
Layout Design: Josh Sisco
Project Coordinator: Stacy Asher

Printed in the
United States of America
Type face: DejaVu
from the DejaVu project on SourceForge

Press Media
(801) 373-6996

First edition

ALSO BY PAUL H. SMITH

Reading the Enemy's Mind:
Inside Star Gate — America's Psychic Espionage Program

Learn Technical Dowsing Home Study Course

Remote Perception: Basic Operational Training course

DEDICATED TO

My father, Fred C Smith, Jr. (1926–2013), who originally suggested the idea that led to this book. . . though he envisioned something closer to a booklet than a volume. Sorry, Dad – I got a little carried away.

CONTENTS

ACKNOWLEDGMENTS

No book worth its salt springs forth in perfection from the fevered mind of its author. For every good book there are wise souls behind the scenes who talk the author down off cliffs of his own creation, or help polish the words, or who simply suggest a better noun here or a verb there. Their assistance yields a much better book in the end. The following people provided sage advice or other sorts of support in the production of this book. My thanks to...

Bruce and Irene Bateman for lending me their dining room as a quiet place to write and escape not just the interruptions of the telephone but the temptations of email. Pam Coronado for her help and advice on the section about forensic remote viewing. Sally Rhine-Feather for the wonderful foreword she contributed. Much thanks for the energy and time that took out of her busy life. Nancy Jeane, for her willingness to be honest and frank about things she didn't like in early versions of the manuscript...and things she did. Brian Josephson and Damien Broderick for insights into quantum physics important to portions of this book, and especially to Damien for some corrective guidance I was wise enough to take. He is blameless for any of his advice I may have been foolish enough not to take. Mark Koltko-Rivera for editing advice after an early reading of the manuscript. Nancy Murphy for some brutal but valuable editing work on part of the manuscript. Leith Tidwell, for the tedious work of copy editing the first printing, and Russell Pickering for finding all the typos I missed in the second one.

Ron Nielson, for his valued advice and help in getting this book printed and ready to go. Jon Noble, for very useful counsel and advice on the entire manuscript from beginning to end. John Stahler, not only for pointing out errors, but for his friendship and frequent much valued sanity checks. Neil Strauss for crucial advice he gave about reordering and reorganizing the content of the book. Gene Troy and John Alexander, for reading the manuscript and for being my friends. Jamie Walker, for useful input on an important chapter. Ed Riordan, Will Smith, Shamra Smith, Daryl Gibson and Jamie Conrad Bates for helping with photos in various chapters.

Also, a special thanks to the design team from the University of Nebraska at Lincoln, consisting of Kris Mangrum, Josh Sisco, Natalie C. Henning and their professor, Stacy Asher. The work they performed in making this book so professional looking and beautiful was not just amazing—but miraculous. Finally, I owe my wife and professional editor Daryl Gibson a great debt of gratitude, not only for the spot-on editing advice she always gives, but for playing devil's advocate to help keep me on an even keel and at least marginally in touch with reality. Her textual eviscerations may be painful, but they are always necessary. If I have missed anyone that has helped me with this book, it was unintentional. I apologize and promise to make it up to you in some way in the future.

FOREWORD

I grew up in a unique household where "remote viewing" of ESP cards or throwing dice to "will" a specific die face to come up was just part of the ordinary routine. You see my parents, Louisa and Joseph Banks Rhine, were the founders of modern parapsychology and my childhood in the 1930's was the most groundbreaking time of their early scientific studies. Now I am continuing their work after retiring from a career in clinical psychology. I even had the honor of serving for a time as the director of the institute they founded, now known as the Rhine Research Center, which is the oldest parapsychology research center in this country and probably in the world.

J.B. and Louisa Rhine, Sally Rhine Feather's parents and founders of modern scientific parapsychology.

A few years ago my brother confessed that as a child he had been upset by people frequently asking him if he was going to follow in his father's footsteps. As a younger sister I was never asked that question but I would have answered an enthusiastic "Yes!" I've always felt proud and positive about my parents' work and recall many pleasant memories. One example was when my third-grade schoolmates and I were invited to participate (and earn rewards) in ESP card guessing tests after school at the old parapsychology lab at Duke University run by my father and mother. I've continued to have nothing but positive feelings about parapsychology over the years, even when in my college years I discovered that it was more of a controversial field than I had realized.

My father, "JB," always felt that ESP was a natural and normal human capacity, even if some of his subjects performed much better than others. He likened it to the capacity for writing poetry or for playing a musical instrument. Back in those days, JB avoided giving specific suggestions for how one might improve ESP scoring, out of fear that his suggestions would inhibit people from discovering their own methods for success.

Today, however, more has been learned about the conditions for better performance, especially in the popular area of parapsychology known as remote viewing. I am delighted to see this book by Dr. Paul H. Smith focus on the particular issue of ways of improving and/or opening up to one's own natural ability. This provides a fresh addition to the study of human consciousness and the fascinating phenomena that surround it. More than any other ESP-related practice that has emerged in the past 50 years, remote viewing promises to be the most democratic of them all. Already there are thousands of people from all walks of life around the world who have learned how to remote view in some degree or other. Reports suggest that nearly anyone who applies the proper remote viewing principles will see some positive success from their effort.

Remote viewing is a popular topic at the Rhine Research Center. We sponsor a monthly remote viewing group for local folks to meet, discuss, and practice their skills. And we have a variety of informed speakers giving public talks and workshops in our broad educational program, often broadcast live or available on video. Paul H. Smith is one of the leading specialists and popular teachers in the remote viewing field who presents at the Rhine. In this new book Paul adds to the knowledge and guidance he has given in his live presentations here at the Rhine Center, so a far wider audience can benefit from his knowledge and expertise.

The Essential Guide to Remote Viewing fills an important gap in the remote viewing literature, in that it introduces the basics of remote viewing to the beginner but also offers insight and guidance to those with extensive familiarity and knowledge about the subject. In addition to what remote viewing is and how it got started, there is an introduction to the science of remote viewing and even a guide to critical thinking about this topic. The latter is sorely needed for both proponents and skeptics alike. Also in this very comprehensive book there is a thorough tutorial on what the remote viewing experience is like and a guide on how to get one's feet wet with very little time or expense. A guideline of how to select a reputable teacher is an added gem. Some may want to consult the book for this information alone.

When asked what ESP means in the larger picture of things, my father took a position that could be called relative dualism where mind and brain seem, at least from what we know now, to be separate things. The mind cannot be explained entirely by the workings of the brain, but instead may be partly "nonphysical," at least in some sense.

Sally Rhine Feather (R) gives an ESP test to daughter Laura in the Rhine laboratory in the early 1960s. (Courtesy Sally Rhine Feather)

In a later chapter of the book Paul includes this position along with other more conventional explanations of how remote viewing fits into the universe, thus allowing the reader to consider the whole picture.

Now nearing the age of 85, I'm devoting most of my remaining time and energy to the Rhine Center's educational programs, assisting in case studies, and continuing the preservation of the legacy my parents have left for future generations. By presenting a more informed and clearer understanding of remote viewing, *The*

Essential Guide to Remote Viewing is an important step in affirming this same legacy by offering new insights about consciousness to members of the general public, whether they are newly arrived or long present in the remote viewing field.

I am cautiously optimistic about the prospects for strong growth in research and education here at the Rhine Research Center and for the future of parapsychology in our ever-changing society. Paul's book is a valuable contribution in bringing that future to pass.

Sally Rhine Feather Ph.D.
Executive Director Emeritus
Rhine Research Center
2741 Campus Walk Avenue, #500
Durham NC 27705
August 26, 2014

Chapter 1
The Mystery of Shipyard 402

America's Cold War enemy was up to something inscrutable, or so it appeared to President Jimmy Carter's senior defense advisers on the National Security Council in 1979. They were puzzled over the most recent reconnaissance images of the Soviet Union's Naval Shipyard 402. America's spy satellites were keeping an eye on a mysterious new warehouse-like edifice, longer than five football fields end-to-end, sitting in this small corner of northern Russia along the White Sea. The satellite photos revealed an immense roof over a structure that appeared long enough to house one jumbo fleet-sized US aircraft carrier plus half of another one, and was wide enough to stow two full Queen Elizabeth II-sized ocean liners side by side. But unlike other buildings in the vicinity, the massive sliding doors in the end of the building that pointed towards the harbor were a good 400 yards from deep water.

Logic would dictate that the vast building must have something to do with naval construction. But why build a ship the size hinted at by the cavernous dimensions of the structure whose main exit was at least a quarter mile from any open water? It smelled like a ruse. Were the Soviets using a shipyard to disguise the construction of some other kind of secret weapon?

A declassified US reconnaissance satellite photo from 1969 showing the Severodvinsk shipyard before the massive building was constructed. (National Reconnaissance Office photo)

If that was the Soviets' plan, it was working. Both American and NATO intelligence agencies had few clues about what was going on inside. So tight was Shipyard 402's security that none of the CIA's human agents had been able to slip into the building. No revealing radio or telephone communications had come from 402, and aerial photos revealed nothing about what was inside.

If, as the NSC's most current speculation had it, the Soviets were building their own full-sized carrier, such a ship could signal the start of a dangerous shift in the Cold War balance of power. Though how a tugboat, much less an immense floating airfield could get to the water across that long stretch of dirt was unclear. And what if something else even more threatening might be taking shape inside?

Then, a US Navy captain on the Security Council staff thought of an unconventional alternative intelligence source that was even more mysterious than the building in Shipyard 402. Perhaps it might help solve the riddle.

Until they were flattened by bulldozers in 1996, two ramshackle buildings, numbered T-2560 and T-2561 occupied a back lot of Fort George G. Meade, Maryland. The 'T' stood for 'temporary,' though the buildings had been standing since World War II. Little more than shacks surrounded by a screen of oak trees, the buildings were staffed by a handful of psychic spies. The code name they were assigned at the time was "Grill Flame," and their orders were to develop and then use a mental tool called remote viewing. Their training in this form of extra-sensory perception was aimed at giving them the ability to pierce the walls of secrecy erected by foreign adversaries.

Building T-2560, the World War II-era "operations building" in which the military psychics did their remote viewing work. (Courtesy Dale Graff)

A mix of soldiers and government-employed civilians made up the program—all of them except a few administrative personnel were career intelligence professionals. Grill Flame was a "special access program," or SAP, whose existence was known to fewer than two hundred people in the entire world. It was to them that the Navy captain from the NSC turned when other intelligence avenues failed to get inside the looming industrial fortress sitting in Shipyard 402.

The Naval captain gave a set of geographic coordinates to Fred "Skip" Atwater, an Army captain who managed training for the psychic spies and coordinated their intelligence missions. No further information was provided to Captain Atwater, who selected one of his remote viewers, a warrant officer named Joe McMoneagle to find out what the NSC needed to know. Together, the two men exited the unit's headquarters in T-2561, crossed the narrow road that ran between Grill Flame's two buildings, and entered the "operations" building, T-2560, where the unconventional work began. Joe entered a darkened room alone, stretched out on a twin bed, and relaxed. A few minutes later Fred entered, sat down at a desk next to the bed, and turned on a dim, red light so he

Capt. Fred (F. Holmes "Skip") Atwater, from about the time the events in this chapter took place. (Courtesy F. Holmes Atwater)

could see the page in front of him while distracting Joe as little as possible.

"It's time now to take a look at the area designated by the geographic coordinates..." Fred said quietly, and then uttered the latitude and longitude figures in barely more than a whisper. "Relax, concentrate, and describe the target to me."[1] There was a pause, then a mumble from Joe. While Atwater took notes, the reclining soldier began haltingly to relate the perceptions that were forming in his mind. McMoneagle slowly described the inside of the giant structure that the NSC was so worried about 4,500 miles away on the other side of the globe.

This book is about who these people were, what they did (including what they found hiding inside Shipyard 402), and what came of this highly secret program. It is also about what this might mean for you...how you yourself can learn more about what they did—and how you can do it too.

If you are willing to try!

1. These are exact words taken from the CIA's declassified official transcripts.

Chapter 2
Unleashing the Military Mind

I call it my "Men in Black" moment. Like the character played by Will Smith in the movie, with almost no information to go on, I had to say yes or no to a mysterious government program that hinted at a dive into the paranormal. That moment came in the summer of 1983, just shortly after I climbed a set of shaky wooden stairs to the front door of a nondescript building I had barely even noticed before.

My presence there was at the invitation of my Fort Meade neighbor, Fred Atwater. Though we lived side by side on the same Army post in Maryland and the sign on his door said "Captain," Atwater never wore a uniform. All I knew was that he worked in a "black" undercover spy program, but despite my own high-level security clearances as an Army intelligence officer, that was all I had been able to discover. A month earlier, Fred had said he thought I might be a good fit for the hush-hush work he was involved with. That was somewhat perplexing, but also intriguing, and within a week or so I found myself taking a battery of psychological tests in the office of the Army Intelligence and Security Command's staff psychologist.

Those tests had since been scored and accepted, and now I was knocking on the flaking green metal door of a dilapidated building labeled T-2561 and wondering what lay on the other side.

The headquarters of the psychic spy unit, Building T-2561. (courtesy, Dale Graff)

The door opened to reveal a short, pleasant-looking woman in civilian clothes, who flashed me a veiled smile and ushered me into a grimy yellow room at the back of the rickety building. She offered me a seat in a wooden chair at the end of a table that looked like it could have been the same World War II vintage as the building itself. On the table in front of me was a paper on which was printed the tightest non-disclosure agreement I had ever encountered in my Army intelligence career. It was mine to read and sign. It said I was to avoid telling anyone that I was even in the Army if they didn't already know, and I couldn't talk to any of my other Army friends about what I would be doing, regardless of their rank or whatever top-secret clearances they might hold.

As soon as my signature was scribbled on the document, a bearded Army captain named Tom McNear calmly issued the following invitation in a modest voice that belied his extraordinary words: "Our mission is to collect intelligence against foreign threats using a parapsychological skill known as remote viewing. We want to know if you'd be interested in volunteering to become a psychic spy?"

Capt. Tom McNear, about the time he was assigned to the Center Lane Program. (courtesy Tom McNear)

During those frozen moments between Tom's invitation and my answer, my thoughts skipped like a stone across water: Though enamored with fictional stories of ESP as a youngster, I had grown to be moderately skeptical and largely inattentive to it as an adult. But, as I was now quite startled to realize, the Army was recruiting people to investigate and practice it. There must be something to psychic functioning after all. In that instant I somehow knew I had been invited along on the adventure of a lifetime. There was no way I was going to turn down the offer. "Yes," I said.

Tom seemed just as surprised as I was myself by the swiftness and certainty of my answer. I had no clue that I would some day be asking other prospective, equally unsuspecting candidates that same question. And I certainly had no idea that I soon would be performing seeming miracles using my otherwise ordinary brain. By the time I emerged a few minutes later from the splinter palace that would be my duty station for the next seven years, the Army slogan current at the time, "Be all you can be" had taken on an entirely new meaning for me.

The most immediate and obvious question on my mind now was "What is remote viewing?" But the other question that begged for an answer was "How did the spit-and-polish, shut-up-and-do-what-you're-told military ever get involved in it?"

As directed, I started wearing civilian clothes and growing my hair out. Soon I began to learn the answers to both of those questions, plus many more new ones that came along.

What is remote viewing?

The easiest answer to the "what is remote viewing" question is to say that it is a form of remote perception. But to understand that answer, the best place to start is with extrasensory perception, better known as ESP. If you are one of the many people who don't believe in extrasensory perception, please withhold judgment until I've had a chance to explain. There is much to understand about ESP, including how wrong the myth is that science and ESP don't mix—because they can. Not only has science never proven ESP to be false, but the scientific evidence for the reality of ESP is stronger today than ever before. (In fact, 26 years after that first astonishing day at Fort Meade, I earned a Ph.D. from the University of

Texas with a dissertation making the case strongly for the scientific reality of ESP.) By the end of this book you will understand why that myth persists that ESP is bogus when it isn't. But first, why do we call it called "ESP"?

The legendary parapsychologist J.B.Rhine adopted the term "extrasensory perception," or ESP, as a name for the class of phenomena he researched for more than 40 years, beginning in the late 1920s at Duke University. The "extra" in "extrasensory perception" means "outside of" or "beyond," as in extracurricular (outside of your regular studies), extraordinary (outside or beyond what is ordinary), or extraterrestrial, (beyond the Earth - although please don't confuse remote viewing with alien abductions or UFO's.)[1] "Sensory" or "sensing" has to do with our physical senses, and "perception" is the process by which we recognize or become aware of the sensations that come to us by way of our physical senses.

J.B. Rhine, father of scientific parapsychology, who coined the term "ESP." (courtesy, Sally Rhine Feather)

1. There is a long-standing popular belief that the "extra" means additional, as in an "extra," or sixth, sense. This comes from a misunderstanding of what ESP is. As I explain here, it is perception that happens without the senses being involved.

People often confuse "sensing" (detecting something with your five senses) and "perception." Though the two processes are related, they are not the same thing. Sensing comes first; our sense organs pick up sounds, odors, textures and temperatures, colors and qualities of light, and flavors from our environment, which create in us raw sensations or experiences. What our senses detect is then passed along quickly to the perception centers in our brains, which go to work processing these sensations. The result is an awareness or a recognition—a perception—of each sensation. These perceptions then become the experience of hearing, smelling, feeling, seeing, and tasting.

Notice the important difference between "sensing" and "perceiving." Sensing is the collection of raw information through your "sense detectors"—eyes, ears, skin, etc. To sense something you must have direct contact with the outside world. But perception is the internal action your brain takes to transform the raw material into information and experience.

Look at it this way: Your computer has input devices—a keyboard, mouse, scanner or internet connection—to bring outside information into it. Once the computer has the information from those input sources, it goes to work converting the input into formats that can be used in higher processes. Think of your brain as a computer, and your senses of hearing, touch, smell, taste, and sight as being input devices. Perceptions are what your brain makes of that input. These perceptions in turn provide much of the raw material for thought.

But your brain and mind can do something your computer can't. Your brain can receive input that doesn't come through the senses. Thus extrasensory literally means "outside," or "beyond," the senses. In the case of ESP your senses don't bring outside experiences and information to the centers of perception in our brains. Instead, something else does, by-passing the senses altogether and going straight to perception.

Since ESP means the ability to perceive things without the help of our physical senses, many scientists and doubters have a hard time believing in it. There is, at least for now, no way they can explain how the experiences that lead to an ESP perception actually "get inside one's head," and this makes them reluctant to take it seriously.

There are a few theories about how ESP works that I'll talk about later. One theory, though, we can rule out right now. Contrary to what movies and television often show, ESP doesn't involve "brain-waves" of any kind. Parapsychologists (scientists who study ESP) have done experiments that block any kind of electro-magnetic signal, including brain-waves, from coming or going in the immediate vicinity of the research subject during the experiment, yet ESP still occurs.

Among other things, parapsychologists working for the military program did this by putting remote viewers inside special enclosures called Faraday cages that shut out most electromagnetic signals. In other experiments they would leave the viewers outside and put the targets inside the cages. It still worked. In partnership with government-supported scientists, one parapsychologist named Stephan Schwartz even went to the extreme of arranging for a deep-diving research submarine. Remote viewers entered the snug confines of the submarine, which then submerged to a depth of hundreds of feet where virtually all outside electromagnetic radiation was shielded away by the electrically-conductive sea water. Even in the absence of electro-magnetic signals, successful ESP effects still occurred. So far, no other satisfactory physical explanation has yet presented itself to fill the gap left by the discredited brain-wave theory. But there have been some tantalizing suggestions that I will get to later.

Many scientists reject the substantial evidence for ESP simply because they can't yet see any way of explaining how it could work. That strikes many folks as an ironically unscientific attitude. It brings to mind those who, according to the apocryphal story, refused to look through Galileo's telescope because they were convinced moons couldn't possibly orbit around planets since everybody knew back then that everything in the universe revolved around the Earth. This attitude is even more like the very real story of medical doctor Ignaz Semmelweis who, in the mid-1840s, discovered that merely disinfecting a physician's hands led to a much lower incidence of fatal infections.

Yet though he had the evidence to back it up, his idea was rejected for decades, leading to thousands of unnecessary deaths, because it didn't make sense in terms of the scientific theories common at the time. If you are coming to this book with the idea that "there is no evidence" for ESP or believe that "ESP is impossible," then you might find the science portion of this book particularly interesting when you get to it, especially chapters 11, 12, and 13.

There are a number of different labels people use as alternatives for ESP. Examples of these are "psychic ability," or "anomalous cognition," or even just simply "psi" (which is the name for the 23rd letter in the Greek alphabet, standing in the ESP community for "something that we can't yet explain") "Psi" is often used as a collective term for anything psychic, including both psychokinesis ("mind-over-matter") and ESP. I find "extrasensory perception" still to be the best and most accurate way of talking about the skill I learned in the Army—remote viewing—and all its cousins.

There are a number of different disciplines under the umbrella of ESP. I'll name just a few. Telepathy is direct communication between your mind and someone else's mind. Then there is presentiment, a word combination of "pre"—or "coming before"—and "sentiment" (a mental impression or emotion). Presentiment, then, is the subtle feeling people sometimes get about something bad that is about to happen.

Precognition is knowledge or awareness of an event that will happen in the future. For example, my friend Shelia Massey told

The deepocean research submersible Taurus, from which remote viewing experiments showed that ESP did not involve "brain waves" or other kinds of electromagnetic radiation. Standing in front of the submarine are Hella Hammid (L) and Ingo Swann (Courtesy, estate of Hella Hammid, ©Tino Hammid, Los Angeles)

Ignaz Semmelweis, whose theories about germ control took decades to become accepted (engraving by Jeno Doby)

me of a precognitive dream her brother had. He dreamed that he was driving through an intersection he always passed on his way to work when another car roared through the intersection and smashed into his. On awakening the next morning, he thought little of the dream. But, as he was on his way to work, he hesitated a few seconds before entering the intersection he had dreamed about the night before, even though the light was green in his favor. Just as he started to go through, a speeding car ran the red light. If he had not paused, he would have been right in the path of the speeder.

Another type of ESP is the feeling someone is staring at you, even when you don't physically observe someone watching. This "sense of being stared at" (more formally known as scopaesthesia) is one form of what has become known as direct mental interaction with living systems, or DMILS. This research discipline examines how some living beings can interact with other living beings without any identifiable physical link between them. And then, of course, there is clairvoyance. French for "clear seeing," clairvoyance is the ability some people seem to have to "see" scenes, objects or events occurring at a great distance or in a sealed package or locked room, even though there is no way for them to physically sense anything at all from those places.

I started out this chapter by saying that remote viewing is a form of remote perception. So how does remote perception fit into the idea of extrasensory perception? You can think of remote perception as being a subset of ESP. To qualify as ESP only requires that the person receive impressions through extrasensory means of any sort. It doesn't matter whether the impressions become conscious or remain in the subconscious, whether the impressions seem to come from another person's mind (telepathy), through some kind of clairvoyance, or by means of some kind of artifact, such as in scrying or psychometry.[2]

I prefer to limit the definition of remote perception to mean the process by which a person develops his or her awareness of a target apparently directly from the target, and not through someone else's mind or the use of some object or artifact as an intermediary. As I said before, remote viewing is a kind of remote perception. And therefore, like other approaches I have mentioned, it is also a kind of ESP.[3] Remote viewing is especially closely related to clairvoyance.

The following is the military definition of remote viewing, taken from the official Defense Intelligence Agency manual for coordinate remote viewing:

"The acquisition and description, by mental means, of information blocked from ordinary perception by distance, shielding or time."

"Information" as used in this definition isn't just facts that the remote viewer suddenly, magically "knows." Information also comes to the viewer in a form that resembles sensory experiences—even if the physical senses can't possibly be involved. In other words, a remote viewer can perceive colors, smells, textures, sounds and even tastes that turn out to be present at the

---

2. Scrying uses crystal balls, smoked mirrors, etc, to focus attention. In psychometry, a perceiver holds an object in his or her hand and tries to discern the object's "history."

3. For a long while now, I've thought that "remote viewing" should really have been called "remote perception." But it got the name early in its development, and it has stuck.

A diagram showing how remote viewing fits into the hierarchy of mental phenomena.

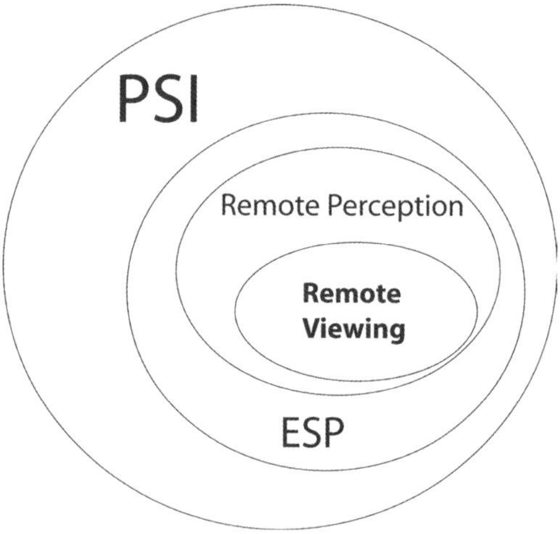

targeted location, all without the viewer having any actual sensory connection with the target. Chapter 8 will paint a picture for you of what the remote reviewing experience itself is like.

As a soldier in 1985 and 1986, I compiled the manual from which the definition I quoted above came. As I was writing it, I borrowed the words of the definition from earlier scientific documents about remote viewing. But I was never completely happy with the definition. As far as it goes, it is correct. Remote viewing does not require direct physical sense perception, technical means (in other words, no television, radio, internet, etc.), hearing about it from someone else, and so on. Real remote viewing information is obtained only through the powers of a viewer's own mind.

The problem is, the definition is not complete. It could describe any of the other forms of "being psychic." If someone uses clairvoyance or a crystal ball or psychic "reading" or even so-called "spirit guides" to gain accurate information about a distant target, it could count as "remote viewing," according to this definition. But none of these things are remote viewing; clairvoyance, scrying, psychometry, channeling and so on have their own names and procedures.[4] So

4. I'm not making any claim about or commitment to the reality of any of these forms of psychic behavior. That's for others to do if they can. My only focus here is on remote viewing and remote perception.

how is remote viewing different from other ways of "being psychic"?

First, even though it is like clairvoyance, remote viewing is different from clairvoyance. Clairvoyance literally means "clear seeing," yet remote viewing involves not just "seeing," but elements relating to other senses as well, such as hearing, touch, smell, and even taste. Some aspects of remote viewing even seem to include telepathic elements. So we could think of remote viewing as a kind of "cocktail" of ESP effects.

Second, the definition doesn't consider the issue of "mental noise." Mental noise is the racket coming from all the thoughts, guesses, imaginings, memories and other background chatter that goes on inside our heads nearly all the time. (If you don't believe you have mental noise, try emptying your mind completely of all thoughts. Bet you can't do it!) Remote viewing differs from all other ESP-based disciplines in that every true remote viewing method recognizes the problem of mental noise and offers some technique or process for dealing with it.

A better definition for the military manual would have been something along these lines:

> **"A process for the acquisition and description, by mental means, of information blocked from ordinary perception by distance, shielding or time...and includes some mechanism or procedure for controlling mental noise."**

One other thing to know about remote viewing; it is not just a method of using ESP. "Remote viewing" also refers to an experimental protocol.[5] In other words, the framework, rules, and procedures within which a remote viewing session is conducted can be considered "remote viewing" as well. There are various remote viewing methods—you can think of "methods" as a set of techniques

---

5. In fact, "remote viewing" originally only referred to the experimental protocol. Applying the term to refer to specific methods came later. There is controversy about the use of the terms "method" and "protocol." In science, the terms are often used interchangeably, and this was frequently so in both the research and the operational parts of the remote viewing program. Noted remote viewer Joe McMoneagle, however, has insisted on this distinction between method and protocol. Since I think it does help make things more clear, I myself now accept this usage, though only conditionally.

for how to do remote viewing. But there is also the remote viewing protocol. Think of protocols as the goals a viewer is supposed to aim for and the conditions under which remote viewing should be done. Take lap swimming as an analogy, for example. There are any number of strokes ("methods") a swimmer can use to get from one end of the pool to the other. But the conditions ("protocols") are the same no matter which stroke is used: there must be water; it must be deep enough; the swimmer must go end-to-end, etc.

Among the conditions for remote viewing, no matter what the method, is that the viewer must be told nothing about the target. We call this condition being "blind" to the target. One good reason for keeping viewers "blind" is that they can't be accused of speculating or outright "cheating." An even better reason to keep the viewer blind is that if a viewer has any idea about the nature of the target (we call this "front loading"), his or her mind instantly fills up with those preconceived notions. This is just another form of "mental noise," and it can easily get in the way of the ESP-based information that comes in on a much more subtle channel. Ironically, knowing what the target is in advance will often lead to the viewer coming up with the wrong information about the target, rather than guessing correct information about it.

A second element of the remote viewing protocol is that no one associated with the viewer (whether before or during the remote viewing session) should have knowledge of the intended target, ei-

A remote viewer working on an RV session.

ther. This is what is known as a "double-blind" condition, and keeps other people from either accidentally or intentionally leaking information about the target to the viewer.

Third, it is important whenever possible that the viewer receive feedback about the target, but only after all remote viewing work is finished and no changes or alterations can be made to the remote viewing results. Feedback is photos, written descriptions, or anything that describes the target to the viewer, so he or she can compare his or her results to ground truth. The viewer can see what he or she got right and wrong about the target—an important learning tool to help the viewer improve. Finally, it is helpful for the viewer to work in a quiet, secluded area where distractions and interruptions are minimized.[6]

These were generally the same conditions under which the remote viewers tried to "see" inside the building at Shipyard 402. Both Joe McMoneagle and his compatriot Hartleigh Trent, who worked as an additional viewer on this project, were "blind" to the target. They didn't know it was in a Soviet shipyard. They didn't even know it was a building. They were taken to a windowless room by someone who also had no knowledge of what the target was. They were given latitude and longitude figures, and asked to describe what was at those coordinates. Only after they successfully described and drew sketches of the building, were they shown satellite photos of its roof, which was all the feedback that was available. At this point they were asked to "go inside" and describe what was happening. And that is what they did.

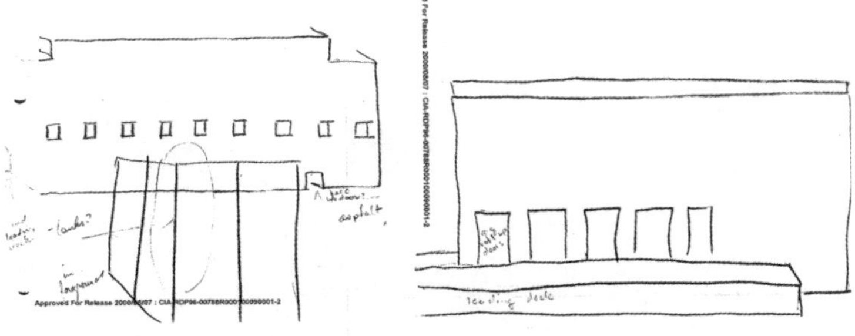

Two of the sketches the remote viewers made of the giant building in Shipyard 402.

6. There are other, more detailed protocol elements, especially when scientific research is intended; but those listed here are the most fundamental.

Chapter 3
The Most Paranormal Thing...

"The most paranormal thing about the remote viewing program was where it got started in the first place." This is how Dr. Harold E. "Hal" Puthoff starts off nearly every explanation he gives of the remote viewing program he founded. He says this because the program got its start in the most unlikely of places, a hard science think tank doing top secret work for the government.

One day in the summer of 1972 two men came knocking on Puthoff's office door. "We're from the CIA," they said. Holding up a document in front of Puthoff, they demanded, "What can you tell us about this?" Puthoff, a pioneering laser physicist working at the Stanford Research Institute (later changed to SRI International) in Menlo Park, California, looked at what they held dangling in the air. It was a paper Puthoff himself had written just a few weeks previously and circulated privately (or so he thought) among a few of his scientist friends.

The paper detailed an experiment Puthoff had conducted a month or so before, in June 1972, with a New York artist and consciousness researcher named Ingo Swann. During the experiment Swann had focused his mind on a large, heavily shielded quantum particle detector in the basement of Stanford University's Varian Physics Building and caused the readouts from the machine to fluctuate wildly. Swann also managed with a fair amount of accuracy to describe and sketch the particle detector's inner structure – details that had never been made public. Swann was using something he called "remote viewing," which Puthoff had never heard of before, to "see" the inside of the device. Now the CIA had come calling, wanting to know what was going on.

The nation's premier spy agency was worried, and rightly so. Over the prior decade US agents abroad had been tracking a massive Soviet investment in what was loosely termed paranormal research. The USSR was looking into a wide variety of psychic behaviors, including psychokinesis ("mind over matter") and clairvoyance, and was spending many millions of dollars on something that most scientists in the West were convinced didn't exist. The Soviets never spent money on that order of magnitude for anything that didn't promise a payoff. Hence, the CIA's worries. Was there really something to this ESP stuff after all?

Hal Puthoff and Ingo Swann in the 1970s. Their first experiment attracted the attention of the CIA and led to founding of the government's psychic spy program. (Courtesy, Hal Puthoff)

DST-1810S-202-78

DEFENSE INTELLIGENCE AGENCY

PARAPHYSICS R&D—WARSAW PACT (U)

PREPARED BY
U.S. AIR FORCE
AIR FORCE SYSTEMS COMMAND
FOREIGN TECHNOLOGY DIVISION

Front cover of a SECRET study documenting the extensive Soviet and East European paranormal research,which fueled US Government interest in remote viewing.

That first visit to Puthoff's door at SRI blossomed into a multi-year government research contract which was eventually passed from the CIA to the Air Force, then the Army, then the Defense Intelligence Agency. Some 23 years later, Congress ordered the remote viewing program, by then known as "Star Gate," returned full circle to the CIA. I've told the history of military remote viewing in more detail in my book *Reading the Enemy's Mind: Inside Star Gate America's Psychic Espionage Program*, so I'll only give a brief sketch here of what happened over the quarter of a century the program existed.

1972-1975: The CIA's Project Scanate

The CIA was impressed enough with Puthoff's credentials and trustworthiness that in late 1972 it gave Stanford Research Institute $50,000 to start a research program with Puthoff in charge. More funding soon followed. Ingo Swann stayed on as part of the project, and these two were soon joined by another scientist, Russell Targ, who became Puthoff's research partner for a decade. Other scientists signed on as well, including nuclear physicist Ed May. Additional remote viewers also became involved – among them a woman named Hella Hammid and a man named Pat Price. Price went on to produce a number of impressive results and was hired away from SRI to work directly for the CIA until his death from a heart attack in 1975. At first, only the scientists themselves and the civilian research subjects with whom they worked were involved. But that was soon to change.

The Military/Operational Side of Remote Viewing
1975-1979: The Air Force

In 1975 the CIA came under condemnation from the US Congress and Senate for a number of misdeeds. (The CIA was dubbed "the rogue elephant" of the intelligence community by Idaho Senator Frank Church. Church was chairman of the Church Committee that investigated the CIA's many illegal and unethical infractions.) William Colby, the new CIA director brought in to "clean up" the Agency, ordered the termination of all illegal or controversial activities. Research into ESP was considered to

(L to R) Hal Puthoff and Russell Targ (codirectors of the remote viewing program), Kit Green (CIA scientist), and star remote viewer Pat Price. (Courtesy Russell Targ)

be controversial, so the remote viewing program, by then called by the code name "Scanate,"[1] was cancelled. Puthoff, Targ, and their associates at SRI were thrown into a tail spin, and for awhile it looked like the remote viewing effort would have to close down.

Fortunately, the advanced research office in the Air Force's Foreign Technology Division (FTD) had become interested in the remote viewing program. Under the leadership of an intelligence analyst named Dale Graff, FTD took over the program and ran it until 1979. The Air Force's taking over the lead for the remote viewing program from the CIA was the beginning of a split in how the US Government pursued remote viewing. This signaled an era where the military would come to establish its own in-house remote viewing ability, in addition to what was being done at SRI. Within a few short years, two related but mostly independent programs emerged. They were still managed under the same overall administrative umbrella, but each focused on a distinct aspect of remote viewing. The military side of it eventually coalesced on the East Coast in the Washington, D.C. area, mainly at Fort Meade, Maryland, and focused on using remote viewing operationally to

1. Short for "scanning by coordinate"—a code name to protect the true nature of the program from being known.

collect intelligence in support of military and intelligence community goals. But it also had a small research mission as well. The second half of the program continued at SRI, and concentrated largely on theoretical and applied research, while also performing some operational work.

Dale Graff, who saved the remote viewing program by arranging for Air Force sponsorship when the CIA backed out in 1975. (Courtesy Dale Graff)

In 1979, Dale Graff moved to the Defense Intelligence Agency, the top intelligence organization in the Department of Defense. The Air Force dropped all involvement in remote viewing research. But Graff joined a small Defense Intelligence Agency project which was also participating with SRI International in looking into ESP phenomena. Some important remote viewing successes happened during the Air Force era. I will mention only one here.

In March 1979, not long before he moved on to DIA, Dale Graff handed a young Air Force enlisted woman named Rosemary Smith a map of the entire continent of Africa. She was told only that sometime in the past few days a Soviet Tu-22 bomber outfitted as a spy plane had crashed somewhere on the continent. The United States desperately wanted to recover the top secret Russian codes and equipment the Tu-22 carried. Using her remote viewing skills, she pinpointed the wreckage, even though it had been completely swallowed by the jungle canopy into which the jet had plunged nose first. Another remote viewer at SRI, Gary Langford, gave an excellent description of the location. Neither the Soviets nor the best American spy satellites had been able to find it, but the information from these two remote viewers led US special operations forces directly to it.

An intelligence photo of a Soviet Tu-22 bomber. A Tu-22, outfitted as a reconnaissance aircraft and lost in Zaire in 1979, was located by an Air Force remote viewer. (Defense Intelligence Agency photo)

Rare public confirmation of this success came in 1996 when former president Jimmy Carter, while speaking that year at Emory University's commencement, was asked what the most unusual event of his presidency had been. Carter replied that CIA director Stansfield Turner had called on a psychic to help in a search for a plane. The psychic then "gave some latitude and longitude figures. We focused our satellite cameras on that point and the plane was there," Carter recalled.[2] Speaking almost 20 years after the fact, he got some of the details wrong. But the gist of his personal testimony confirmed the reality of the remote viewing success.

2. "Psychic helped locate downed U.S. plane, ex-president says," Reuters, September 21, 1995. (President Carter didn't specify, and the reporter didn't know, that the aircraft was Soviet, not American.)

1977-1983: The "Grill Flame" Period

In 1977 the US Army Intelligence and Security Command began its own remote viewing program, which in 1979 became known by the code name "Grill Flame"—a name the DIA program soon adopted as well in partnership with the Army. Under direction from Major General Ed Thompson, the Army's senior intelligence officer, then-Lieutenant Fred "Skip" Atwater began the program, together with Major Murray "Scotty" Watt. Watt eventually left, to be replaced over time by a series of other program directors. But Atwater shepherded the program for 10 years, guiding and mentoring many of the most prominent military remote viewers known to the public today. Among these were Joe McMoneagle, Mel Riley, Lyn Buchanan, William G. "Bill" Ray, and myself, plus many more who are not so well known or remain unknown to the public.

Grill Flame's creation ushered in a confusing period during which there were parallel remote viewing efforts going on in the US government. One was the Army program, located at Fort George G. Meade, Maryland, which quickly came to focus on using remote viewing for active intelligence work. Another was the office at the DIA, headed by Dr. Jack Vorona, DIA's top scientist, and managed by Dale Graff. There was also the ongoing work at SRI in California, sponsored jointly by the Army's program and the Defense

F. Holmes "Skip" Atwater, at one of his rare appearances in uniform. (Courtesy F. Holmes Atwater)

Intelligence Agency.[3] Hal Puthoff and Russel Targ still led the SRI effort, which focused on research while doing some operational work.

During the Grill Flame days, newly minted Army remote viewers, drawn from a group of experienced Army intelligence NCOs, officers, and civilians were sent to the SRI lab for testing and orientation in the remote viewing protocol. Among these were Joe McMoneagle, Mel Riley, Hartleigh Trent, Ken Bell, and other pioneer Army viewers. Though what they did for their few weeks at SRI was called "training," it was really more of a familiarization, as no real training method had yet been developed. The new viewers were left to develop their own techniques through trial and error, helped along by Atwater's guidance as he came up with ideas and suggestions for applying remote viewing to problems unique to military intelligence needs.

The Army remote viewers did their first live remote viewing sessions in September 1979, and enjoyed some immediate success. It was during this time, 1979 to 1983, when some of the more spectacular remote viewing successes occurred. Among these was the building in Shipyard 402.

1983-1986: "Center Lane"

I joined the remote viewing program as a junior Army captain at the end of the summer of 1983, about the time the program's code name was changed from "Grill Flame" to "Center Lane." This was just in time for me to enter a new training program contracted by the Army from SRI. Puthoff and Swann had worked together to create a method to teach a structured remote viewing skill called 'CRV' for "coordinate remote viewing." Tom McNear, the bearded captain who had been the one to initiate me into the world of psychic espionage, was the first military guinea pig to go through the CRV training beginning in 1982.

The new training contract began in January 1984. At first there were to be three of us—Capt. Bill Ray, an Army civilian named Charlene Shufelt, and myself. Shortly before training was to start a

3. Yet another remote viewing program, sponsored at Aberdeen Proving Ground in Maryland by the Army's Materiel Systems Analysis Activity, lasted only a couple of years, but made a number of valuable contributions.

fourth person was added, Capt. Ed Dames, who worked for another intelligence unit on Fort Meade. Our training required numerous weeks away from Ft. Meade at the SRI offices in California and in New York City, with Hal Puthoff, but mostly Ingo Swann as our guides and instructors. Our training program with SRI wrapped up in December 1984, just in time for us to deal with a growing bureaucratic crisis over the future of remote viewing in the military.

Ingo Swann with his students in 1984. (L to R) Bill Ray, Paul H. Smith, Ed Dames, Ingo Swann, Tom McNear and Charlene Shufelt.

Power struggles in the Army forced the transfer of the remote viewing program to the Defense Intelligence Agency, where the Army's operational intelligence program was swallowed up by the DIA's own effort. The transfer process began in 1985 and ended officially on January 31, 1986. All along, DIA had been the primary manager of the work SRI was still doing in California, so finally all military remote viewing efforts were together under one headquarters at the Defense Intelligence Agency.

1986-1990: The "Sun Streak" Era

With the move to DIA, we acquired a new code name, "Sun Streak," and more personnel were assigned to us permanently—among them Ed Dames and Mel Riley who still today enjoys some fame in remote viewing circles. Sun Streak introduced a "golden age" of sorts for

Star Gate military remote viewing personnel in 1985: (L to R) Paul H. Smith, Lyn Buchanan, Bill Ray, Fred Atwater, and branch chief Lieut. Col. Brian Buzby.

military remote viewing. We had reliable funding and sponsorship and were introduced to a wide array of new targets. These ran the gamut of intelligence work: American servicemen that might possibly still be held captive in Southeast Asia (we found none); new Soviet weapons technology; Mideast hostage situations; the location of Iranian Silkworm missile sites; finding CIA moles; the Pakistani nuclear program. Eventually, one of our biggest customers became the military's counter-narcotics community. Starting in 1988 and lasting into the early 1990s the remote viewers of Project Sun Streak joined the war on drugs. We worked many sessions against drug transport ships, transhipment points, and production facilities.

We had our most successful time in this mission during 1990, when the Army intelligence staff in the Pentagon "borrowed" Sun Streak from the DIA to support the Army's role in the counter narcotics joint task forces in the Caribbean and Pacific. In evaluating the work that we did then, the task forces reported that remote viewers were more successful than any other intelligence discipline in providing usable intelligence on the targets they were assigned. One senior intelligence staff officer observed that "Contraband was recovered and traffickers captured based solely on remote viewing data."[4]

Unfortunately, forces were already gathering to challenge the existence of the military remote viewing program for the last time.

4. Col. William P. Johnson, Office of the Deputy Chief of Staff for Intelligence, 1990. You can read more details about this in my book *Reading the Enemy's Mind*.

1990-1995: "Star Gate"

In late 1990 the program's code name was once again changed, this time to "Star Gate."[5] Counter narcotics operations continued, along with support of the Desert Storm war against Iraq, as well as a variety of other missions—among them attempts to ferret out tunnels being dug from North into South Korea. On the research side, Russell Targ had left the program long before, in 1982. Hal Puthoff continued as director until 1985, when he was offered his own laboratory, the Institute for Advanced Studies, in Austin, Texas.

Dale Graff with remote viewing research personnel in the early 1990s: (L to R) Dale Graff; researcher Nevin Lanz, Ph.D.; former Star Gate remote viewer Ken Bell; and Joe McMoneagle. (Courtesy Dale Graff)

Ed May became director of the SRI program, which remained at SRI International until 1990, when May negotiated a transfer to another company, Science Applications International Corp. (SAIC), and began to work more closely with the military operational viewers in some of his research projects.

Unfortunately, military politics again began causing mischief. Despite Star Gate's successes, high ranking officials in the Defense Intelligence Agency disliked having to support the remote viewing effort, even though the program had the backing of prominent

5. Bearing no connection to the 1993 movie or later TV series of the same name.

Dr. Ed May, a nuclear physicist who joined the SRI remote viewing laboratory in 1976, became its director and chief scientist with the departure of Hal Puthoff in 1985. (Courtesy, Ed May)

members of the House and Senate. The DIA leadership made it an unofficial policy to kill the program by attrition—transferring viewers and other personnel out of the program without replacing them. A move was finally made to take Star Gate away from DIA and give it back to the CIA. Unfortunately, the CIA was then headed by director John Deutch who had a history of objecting to the idea of remote viewing. The Agency decided to cancel the Star Gate program once it had official control of it.

On June 30, 1995, the very day that Congress transferred the program to the CIA, the Agency shut down Star Gate and scattered its personnel among the various intelligence organizations in the Washington, D.C., area. With the subsequent publication of a report declaring that remote viewing was of "no use" as an intelligence tool, remote viewing was effectively dead in the United States intelligence and military communities.

1995 to the present: Remote viewing out in the open

On the evening of November 30, 1995, Ted Koppel's *Nightline* television program featured a surprising topic. Americans, most for the first time, learned that their government had been directly involved in exploring and using ESP in defense of national security. On the show was Robert Gates, a former CIA director and future secretary of Defense; Dale Graff, now retired after eighteen years as part of the government remote viewing program; Joe McMoneagle, one of the star military remote viewers; and a US Customs officer still on active duty who called himself "Norm." On the air, the Customs officer gave details about one of the more spectacular remote viewing success stories, in which remote viewer Angela Ford provided detailed information leading directly to the capture of a renegade Customs agent who had fled after being convicted of aiding drug traffickers. Gates claimed that as far as he knew, remote viewing had never been of any use. The Customs officer and the others on the show begged to differ.

Dale Graff with Senator Claiborne Pell outside Building T2561 in the early 1990s. (Courtesy Dale Graff)

The CIA's closing of Star Gate may have been one of the best things to happen for remote viewing. When the government stepped out of the picture, remote viewing was no longer hidden behind a wall of secrecy. And that meant that American taxpayers, who for 23 years had funded remote viewing's development, could now themselves explore the practice and enjoy its benefits.

But remote viewing's path into public awareness was not without its stumbling blocks. Some of the same people who tried to turn it into a household word also tainted it with sensationalistic stories and claims, often without any foundation in fact. The hype made it hard for scientists and other critical thinkers to take remote viewing seriously, which in turn made it harder for remote viewing to be taken seriously by the general population. Still, remote viewing has managed to progress. Thousands of people have learned basic

The founding meeting of the International Remote Viewing Association in March, 1999.

remote viewing skills from a bevy of instructors who teach commercially. Remote viewing has been featured in movies and on television.[6] Numerous books have been produced by the original military remote viewers, as well as by outside observers, and many of us have been interviewed by local, national and international media. There is even a nonprofit organization to promote the discipline. The International Remote Viewing Association (IRVA) was founded in 1999. It exists to represent remote viewing in a balanced manner, as much as possible away from the hype and exaggeration.

IRVA hosts regular conferences, maintains several informative websites, keeps a library of scientific resources and articles, and sponsors the Warcollier Prize, (named for René Warcollier, the noted French parapsychologist), an award for the best, most scientifically sound remote viewing research proposal submitted annually.

6. A few examples: an episode of *The Dead Zone*; mentions in the *X-Files* and other series; movies such as *Suspect Zero*, starring Sir Ben Kingsley, Aaron Eckhardt, and Carrie Ann Moss; and, with a great deal of artistic license, *The Men Who Stare at Goats*, featuring George Clooney, Ewan McGregor, and Jeff Bridges.

Today there are many who deserve credit for moving remote viewing along. But those deserving the most praise are Hal Puthoff, Ingo Swann, Russell Targ, Ed May, and Skip Atwater. Without them, remote viewing as we know it might never have existed.

Trying to find out what was inside the building at Shipyard 402 on the White Sea in what was then the Soviet Union, Joe McMoneagle and Hartleigh Trent performed 10 remote viewing sessions over the course of a little less than two months in the fall of 1979. The final session took place on the first of November, when no one in the West yet had an inkling what was going on in the building.

What Joe and Hartleigh described, though, made no sense at all and couldn't possibly be true. They reported that the Russians were building submarines in the Shipyard 402 building. That in itself wasn't too hard to believe; the building was in a shipyard, after all. But what about all that land between the building and the water? And the kind of submarine Joe described was nothing short of fantasy. By Joe's reckoning, this alleged sub was massive—perhaps twice the size of any submarine built to that time by either the USSR or the USA. Not only that, Joe described it as a missile sub. And the missile tubes were in front of the sail or "conning tower," the main superstructure that sticks up from the top of a submarine's hull. This was heresy in naval architecture circles, where it was a matter of principle that the tubes always went on the rear half of the boat, behind the sail.

The Navy captain who took the remote viewers' data back to the National Security Council could get no one to take it seriously. Reportedly, one particular NSC staffer absolutely refused to consider any information coming from such a bizarre, untested source. His name was Robert Gates.

So that is where things stood for nearly 11 more months. And now, the final chapter in the story of Shipyard 402 vs. the psychics.

Chapter 4
The Psychics Were Right

Joe McMoneagle and Hartleigh Trent had accurately described the giant building in Shipyard 402 in reasonable detail. They had revealed even more about its interior, claiming that the immense building held a huge submarine, a radical design far beyond the scale of any submarine ever constructed before. Their data was rejected as being completely unbelievable and unrealistic. And that was where the verdict stayed for almost a year.

Eventually, the Soviets started digging. They excavated a massive volume of dirt on the west end of the building, a hole stretching nearly a quarter of a mile to the sea, almost three football fields wide and perhaps a dozen stories deep, and let in the sea to fill it up. On September 27, 1980, almost 11 months after the final remote viewing session against the Shipyard 402 target, the Soviets floated out the TK208, the first of the *Akula* type super submarines, known today in Western circles as the *Typhoon* submarine class. The sub was longer than the Washington Monument is tall, and nearly as big around as any two US submarines put together. It had 20 tubes holding huge multi-warhead R39 submarine launched ballistic missiles in a double row on the forward part of the hull—in front of the sail. The *Typhoon* caused instant consternation in US intelligence circles. It was a tactical and strategic surprise to the entire Western world. This new class of ICBM-toting subs promised to be a bigger threat than a whole fleet of Soviet aircraft carriers.

At the time, I was serving as the strategic intelligence officer for the Special Forces command in Germany and had not yet even so much as heard the term "remote viewing." It was my job to review and make available to the commanding officer, Col. Bill Holt, the latest top secret intelligence obtained by the Special Security Office in Munich. I still recall the almost panicked tone of the intelligence cables and briefs I read and passed on to the colonel. The Soviets had outmaneuvered the Western intelligence services, taking them completely by surprise, and both sides knew it. With the advent of the *Typhoon/Akula* class submarine, the world had suddenly become a much more dangerous place. And, as I was to learn just two years later, the legitimacy of remote viewing as a useful intelligence collection tool was once more confirmed.

The story of the remote viewing of the Typhoon is not hearsay. It is not just another old "war story" that my fellow Star Gate remote viewers and I expect you to believe just on our say-so. The provenance (the chain of custody) for this information is impeccable. Copies of nearly all the original documents that still exist became available when the CIA publicly released the Star Gate project's archives in 2004. Until then they had been in the protective custody of the CIA. The account I have given in this book of the remote viewing of the *Typhoon* is authentic, taken from the actual records and not from anyone's faulty memory.

If the members of the National Security Council had been paying attention, the remote viewers could have told them almost a year in advance of that September 1980 surprise what was really happening inside the mysterious building in Shipyard 402. Over the next few pages you will see the visual evidence of just how well the military psychics did.

The following two sketches resulted when a geographic coordinate was read to McMoneagle by Atwater during the remote viewing sessions in 1979. Note the industrial appearance of the structures depicted.

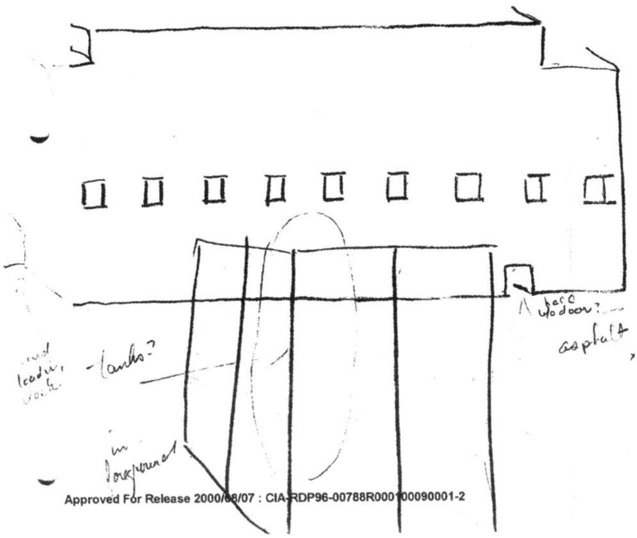

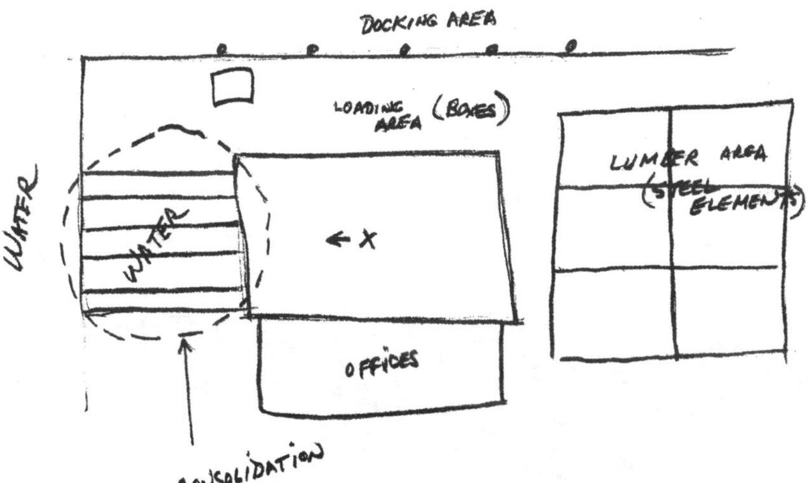

One of the viewers made this overhead sketch of Shipyard 402 and its immediate vicinity. Note the relationship between structures, water, and surrounding area.

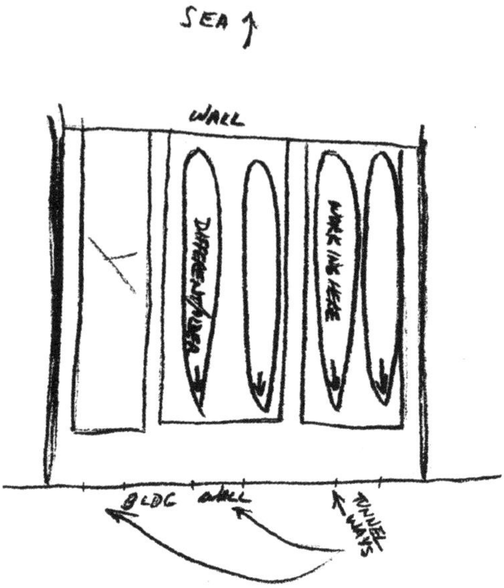

You can see clear conceptual representation of ship construction bays inside the massive building. Also note that the sketch above shows more than one vessel under construction in each berth.

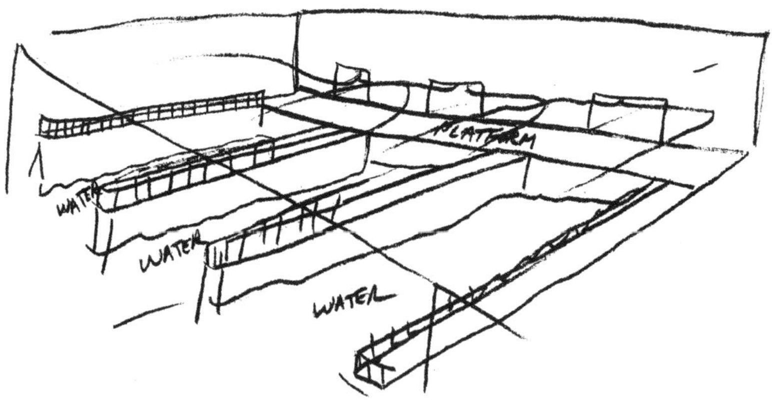

One viewer observed and sketched a huge overhead traveling crane.

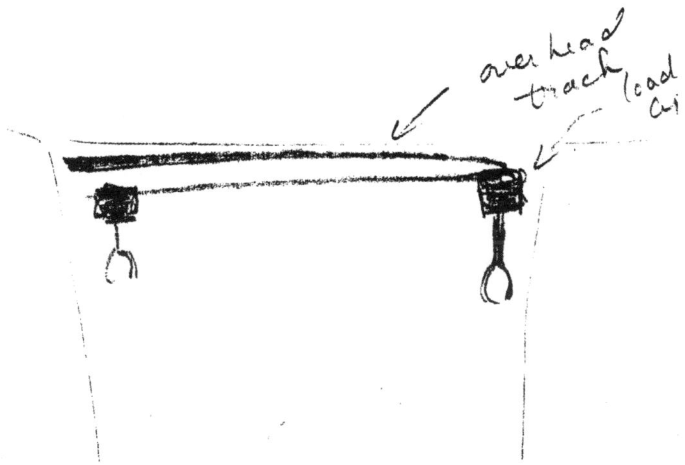

The sketches on the next page provide more insight into what was going on inside 402. The first sketch, along with the verbiage the viewer used to describe the setting indicated that there were multiple subs of varying sizes present in the construction bays. The sketch below that shows conceptually how one of the subs was berthed in its construction bay.

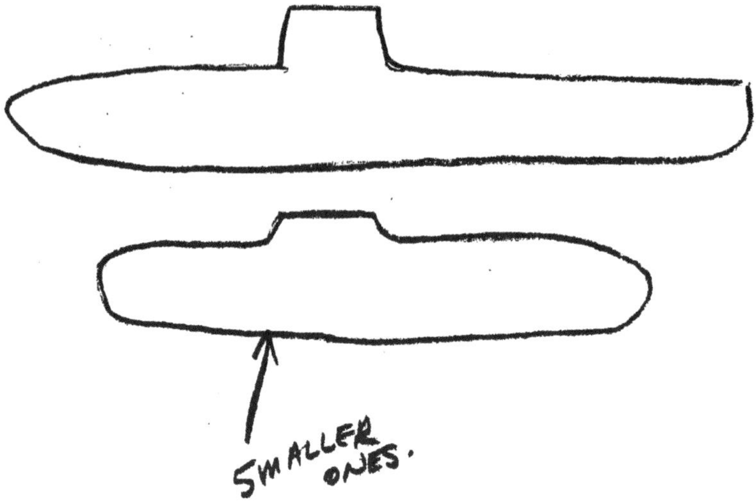
SMALLER ONES.

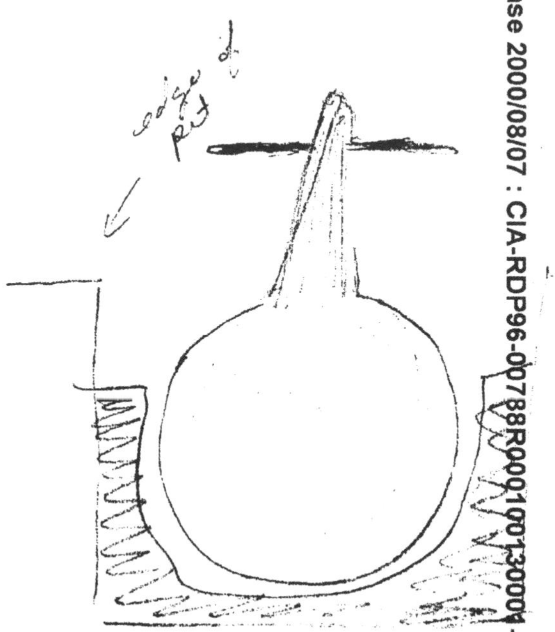
edge of pit

PAUSE

#1 There's a large ship ~~on the water~~ by the wharf. A very large ship. It's got a hollow area in it. And, there's a big...like a...a...a very large submarine. It's there by the big ship, and there's all kinds of hoses going to it, or electric lines going to it. Some kind of lines. There's a...there's a lot of activity on that outside of the building. I don't see anything other than that.

#66 Alright. Tell me more about this activity in the wharf area.

#1 They're working on the submarines.

#66 And where is the submarine?

#1 Between the wharf and the large ship.

The excerpt above from the transcript of Joe McMoneagle's verbal description emphasizes the immense size of the submarine he was perceiving, as does the accompanying sketch (below) of the submarine moored next to a submarine tender vessel.

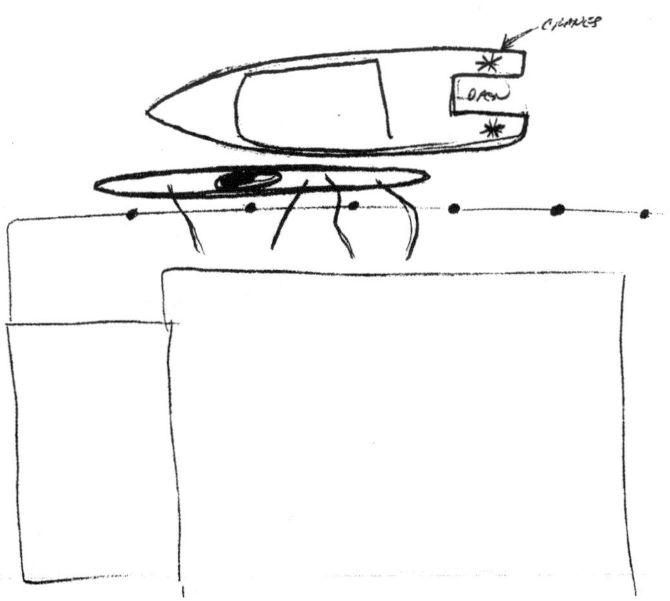

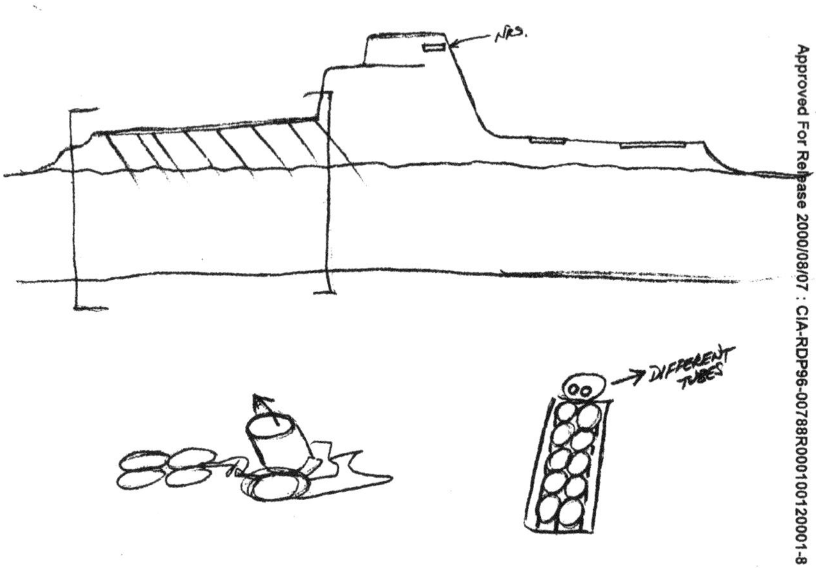

Here is Joe McMoneagle's side view of the submarine he perceived during the remote viewing process. Note his depiction of the missile tubes in front of the submarine's sail (conning tower). The smaller sketches are meant to give an idea of how the missile-tubes appeared from different angles. In the completed Typhoon the tubes were covered by rectangular doors. At the time Joe perceived the sub, it was still under construction and the doors likely had not yet been added.

The preceding gives you a good picture of what the remote viewers were perceiving before anything was ever known to outsiders about the contents of the building in 402. Now I will include photos of the actual remote viewing target compared with the viewers' sketches so you can compare what the viewers produced to the reality revealed almost a year after the final viewing was done.

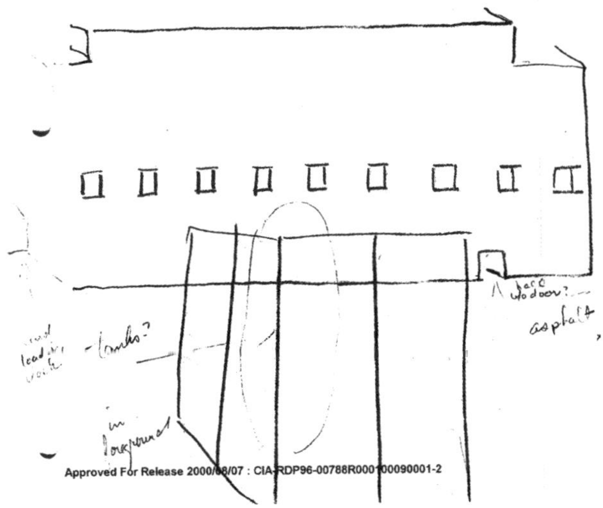

(Sevmash photo)

As you can see, while not capturing every detail, the viewers' sketches nevertheless recorded a generally accurate concept of the structure they were assigned to remote view.

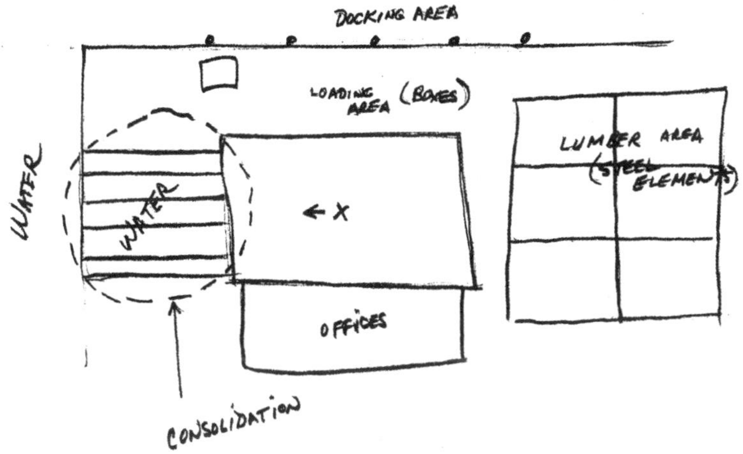

(Google Earth photo)

Notice the general placement and orientation of the viewer sketches as compared to a modern satellite view of Shipyard 402 (now called the Sevmash shipyard) and its vicinity. Though many changes have been made to the facility over the years, the general layout remains essentially the same today.

(Sevmash photos)

The photograph at top is from an early phase in the original *Typhoon's* construction. The photo immediately above is from modern day construction and repair activities.

As you can see from the comparison below, the conceptual sketches of the viewers substantially match the actual interior layout of the immense building. Interestingly, even today the construction births may be occupied by more than one vessel side by side.

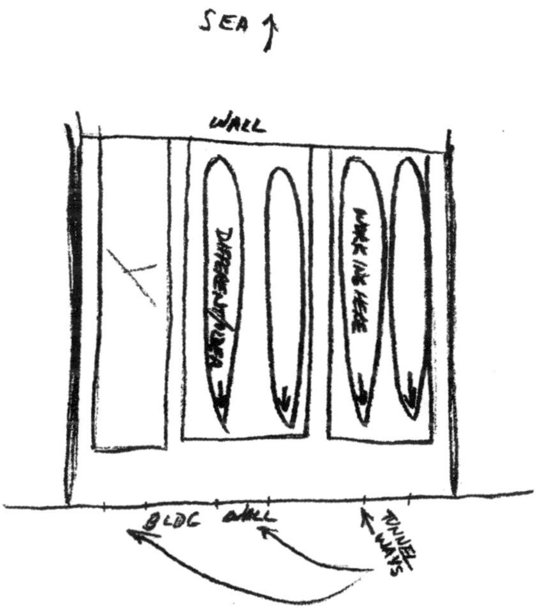

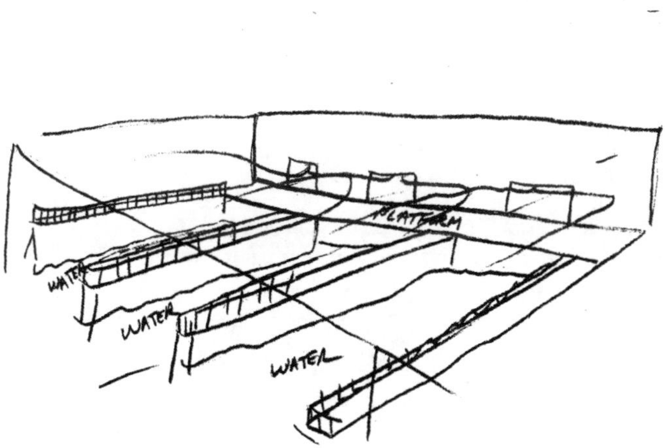

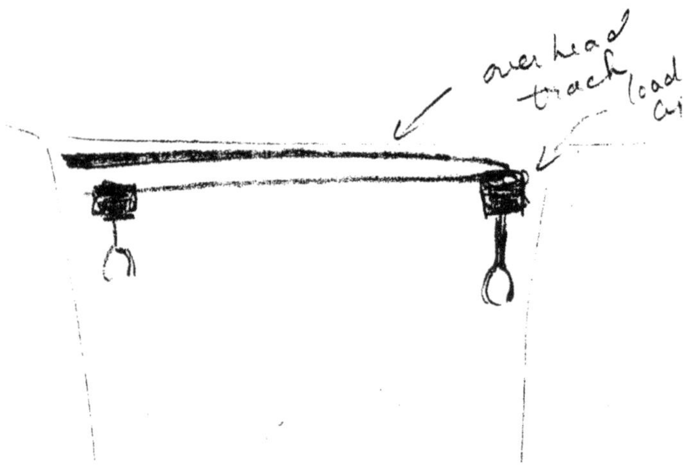

(Sevmash photo)

The photo and sketch above depict the immense, two-hook traveling crane carrying a huge girder during present-day ship construction inside the building. While one would expect to find large cranes present in a facility like this, there are many ways such a crane might be configured. In this case the remote viewer got it exactly right.

(Sevmash photo)

Compare the photo above of a Typhoon--class sub in the final stages of construction with the viewer's rough sketch below.

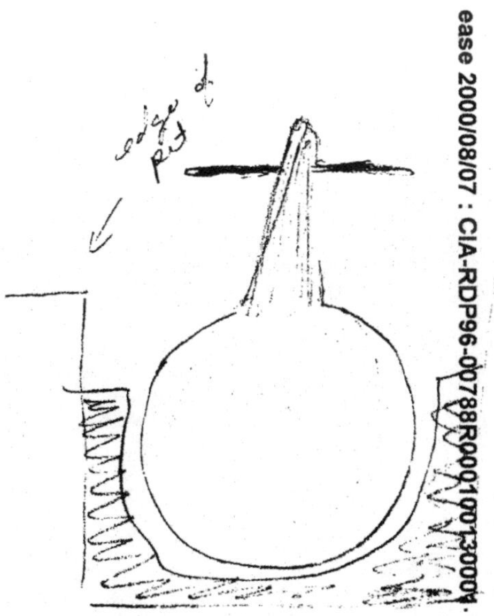

Around the end of September 1980, satellite photos of the area showed a gigantic submarine tied off at the quay next to 402 undergoing the final stages of completion. This submarine was given the official NATO designation *Typhoon*, though the Soviets called it the *Akula* class, or Project 941. In the above photo from 2012, you can see left of center a *Typhoon* class sub tied up next to Building 402. For scale, just to the right of the sub is the full sized aircraft carrier Russia was transferring to India. Both are dwarfed by the nearby building where the submarine began its life.

The image at the bottom is a Defense Intelligence Agency artist's rendering from satellite photos of a submarine tender re-arming a *Typhoon* sub. It is interesting to compare the DIA artist's depiction with Joe McMoneagle's remote viewing sketch from 1979 of a similar vessel alongside the *Typhoon*, before the DIA ever knew the submarine existed.

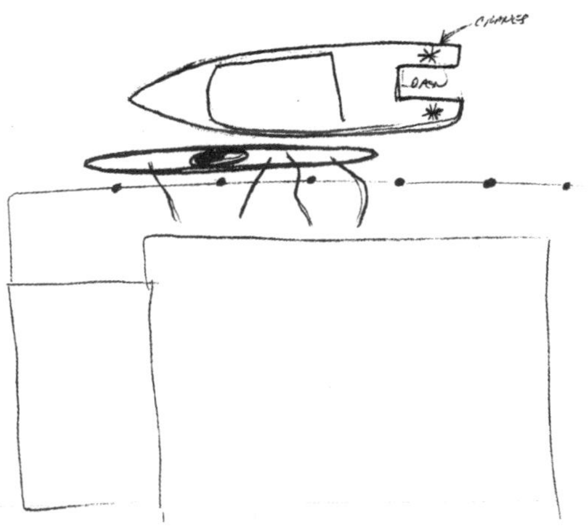

(Defense Intelligence Agency image)

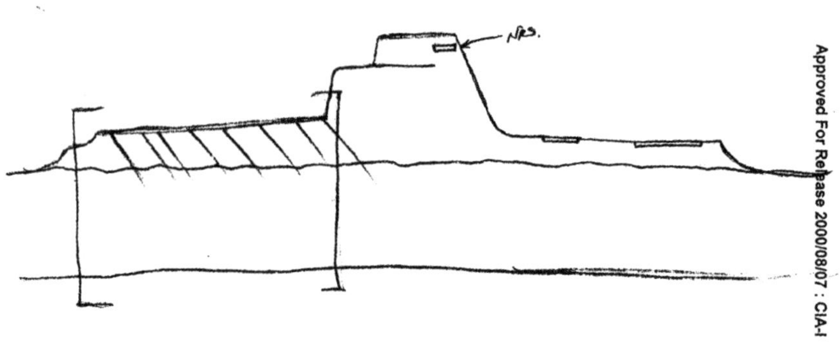

Compare Joe McMoneagle's October 1979 sketch (above) showing the side view of the submarine he was perceiving inside 402 with a photo of the actual sub (right). Note the missile tubes that he correctly located in front of the submarine's sail. Note also the similarity between the sail of the Typhoon and McMoneagle's sketch. They are clearly not identical, but there is a noticeable family resemblance.

There are other details that aren't precisely correct. For one thing, McMoneagle was convinced that the missile tubes were slanted forward at an angle, but in reality the tubes are vertical. Still, small inconsistencies like this are typical of witness reports of events or locations perceived under less than ideal circumstances. Considering that McMoneagle and Trent were having to gain their information from a mental channel known to be of limited bandwidth and with well known signal to noise issues, it would be truly surprising were there not some small differences between their perceptions and the actual target.

The story about Building 402 and the *Typhoon* is only the first of many success stories from the annals of military remote viewing. Today, the military doesn't have the monopoly on the exciting possibilities for what can be done with remote viewing. The next chapter examines many of these possibilities—some of which you might even want to explore for yourself.

Chapter 5
Putting Remote Viewing to Work

If you're like most people, just about now you're saying, "This is all very fascinating, but what is remote viewing actually good for?" Remote viewers far and wide are still in the process of answering this question, and finding rich ground as they do. There are several practical uses, which I will address in this chapter, but the "use" that seems to draw the majority of students to learn remote viewing is this: They want to understand more about themselves, about who we as humans really are, and what more we are capable of than we may have imagined.

The reality of mainstream education in the 21st Century is that it focuses on career development, making a living, and staying in step with economic demands. So when I tell people I teach remote viewing, I am almost always asked, "Why would anyone take your classes?" But what they're really saying, and what you may have asked yourself is, "Why would anyone pay for that kind of class?" or "What can you actually do with remote viewing?"

While many new students have practical uses in mind, most of those who take instruction from me also see learning to remote view as a way of achieving some level of what is often referred to as "self-actualization." Popularized by psychologist Abraham Maslow, self-actualization has come to be almost a cliché, but it does reflect something deeper that any healthy person wants—to achieve his or her full potential as a human being.

Once men and women discover that this "full potential" doesn't stop with their connection to the physical world—that consciousness can extend far beyond the confines of the brain—they feel the urge to explore those wider dimensions of their potential as well. I am always gratified to watch as a new remote viewer realizes that he or she has just received real impressions and experiences that could not possibly have come through either his or her imagination or the physical senses. It can be life changing.

Of course, when we want to know the "uses" for something, we are usually interested in practical applications, not just personal transformation. And remote viewing has a variety of those to consider.

Intelligence Gathering

First on the list is the use to which remote viewing was put in the military: to collect intelligence. In previous chapters I gave you a few examples of this kind of work. Unfortunately, there are far more than I can go into here, though you will read of a few more later in the book. In the military, remote viewing was usually called upon when all other means to solve a vexing intelligence problem had been tried and failed. And, frankly, even remote viewing didn't (and doesn't) always succeed. But any intelligence discipline has its share of failures, so remote reviewing is not alone in this.

Just as with any other discipline, you have to understand the strengths and weaknesses of remote viewing to properly use it. You need to know what it can do and what it can't, and the rules for putting it into practice correctly. Unfortunately, supporters and skeptics alike often fail to do their homework, and load expectations onto remote viewing that it was never designed to meet.[1] On the other hand, when those applying it know what they're doing, there is no reason why successes similar to Building 402 and the many others that are now a matter of public record cannot be replicated.

Business Intelligence

Probably the use of remote viewing that most closely matches the way it was applied in the intelligence arena involves business, or "competitive" intelligence. Although the business world is not yet at the point of embracing remote viewing as a standard information-development tool, there are forward-thinking companies that are willing to see what remote viewing can bring to their table. These companies remain cautious about giving away business strategies

---

1. As just one example, perhaps the most common demand from press, skeptics, and even sympathizers is to come up with winning lottery numbers. And when such numbers aren't forthcoming, that is taken as evidence that remote viewing "doesn't work." But legitimate remote viewers have never claimed to be able to do this. In fact, for reasons I briefly discuss in Footnote 3, on page 74, the scientists researching remote viewing established long ago that the methodology was unsuitable for perceiving numbers or even proper names of people and locations. Instead, it has other strengths and abilities.

and goals, and often are reluctant to reveal their involvement with ESP. So it is rare that remote viewing professionals can give any details about the work being done in that arena. However, over the years I and a number of my colleagues have been commissioned by various companies and corporations to carry out competitive-intelligence type remote viewing projects.

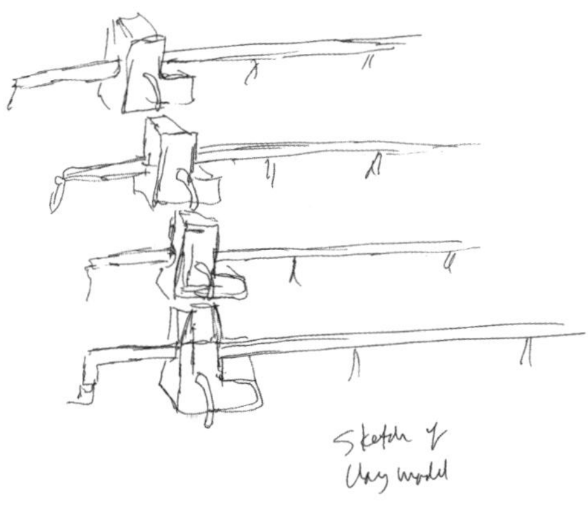

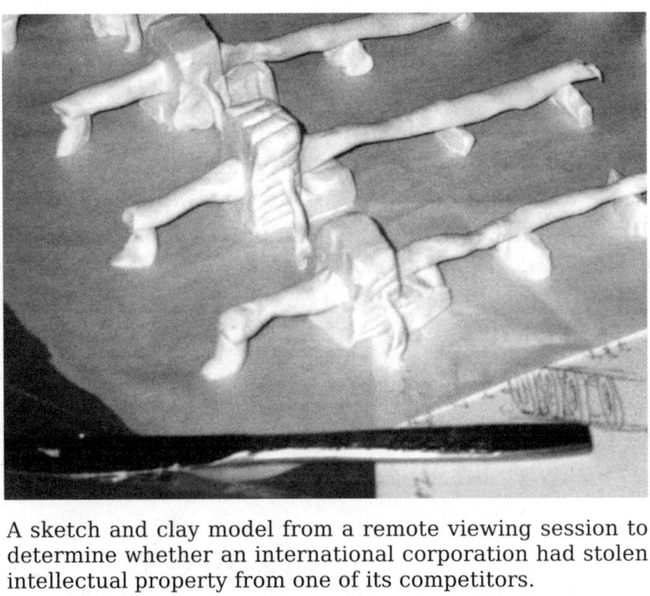

A sketch and clay model from a remote viewing session to determine whether an international corporation had stolen intellectual property from one of its competitors.

In one case, I and my team were asked to help discover whether a competitor had stolen manufacturing technology from the client's company. In another case, a different company asked us to help find a corporate "mole" among its employees who was stealing proprietary information.

Though they are unwilling to give details, both my former colleagues Joe McMoneagle and Lyn Buchanan have admitted to contracting out their remote viewing expertise. And such activity is not restricted to just the United States. As of this writing, a company in France, IRIS-IC, headed by one of my advanced students, Alexis Champion, has not only organized successful remote viewing conferences and conducted seminars in Europe, but is heavily engaged in direct commercial use of remote viewing in support of client businesses in Europe and elsewhere in the world.[2]

"Forensic" Remote Viewing

Police work and criminal investigation, which I call "forensic" remote viewing, is another obvious area for remote viewing applications. Such activities are often not that much different from the intelligence work that was remote viewing's bread and butter in the military. In either case the objective is to increase knowledge about an event, activity, person, location, or object about which not enough is yet known. From that perspective, remote viewing can be a force-multiplier for law enforcement.

My associates and I have occasionally taken part in this kind of remote viewing activity when requested by the authorities. One case that I can now mention in public took place in 1997, shortly after I retired from the Army and set up my commercial remote viewing company. A sergeant in the King County sheriff's department in Washington state was accused of murdering someone, but after several weeks the victim's body had not been recovered. We were asked to use our remote viewing skills to try to locate the body. I and two of my former military colleagues did a half-dozen sessions, trying to describe the location and setting where the body could be located. We passed the results to our contacts in the sheriff's department.

---

2. Of course, it goes without saying that the same ethical principles apply to using remote viewing as to any other competitive intelligence tool.

A few days after receiving our data, deputies had made plans to inspect some possible sites that matched the description we had provided. At 12 noon, literally to the exact moment they were setting out, they received a phone call.

A sketch from a remote viewing session attempting to locate a murder victim's body, compared to photos of the setting taken from the spot where the body was located. (Photos courtesy Ferenc Zana)

Three hikers had stumbled across the body hidden near a culvert in a wooded, isolated part of the county. Though remote viewing ended up playing no role in the discovery, the case at least was solved. But there was one benefit: Once the body's exact location was known, one of the sheriff's deputies was able to make a thorough comparison of our data to the spot where the body was discovered. He sent us a report with maps, diagrams and photographs.

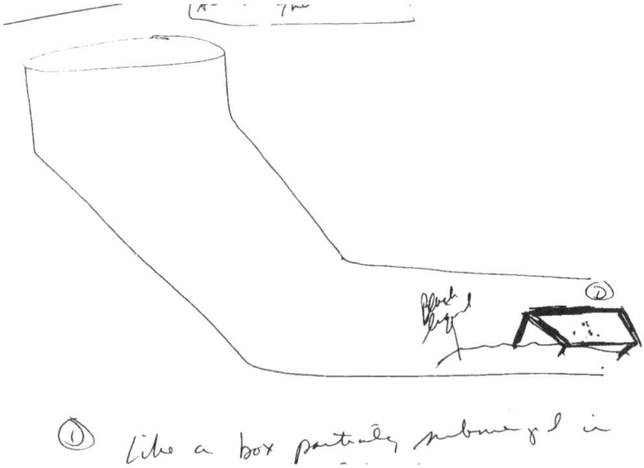

It turned out that our data was surprisingly accurate, describing an abandoned air shaft with a grate over it that was nearby, a creek and culvert closely matched those located within a few feet of where the body was found, as well as other features and details that three former military remote viewers more than 2,000 miles away on the East Coast could not possibly have been aware of.

Sketches from another remote viewing session attempting to locate a murder victim, showing a metal grating covering a mineshaft in the same park where the body was located. (Photos courtesy Ferenc Zana)

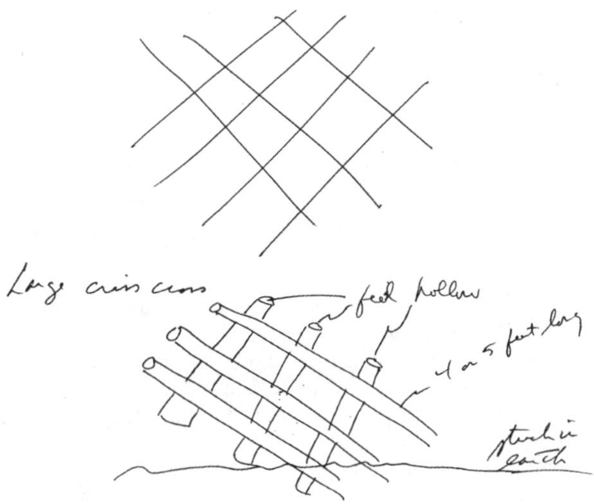

The "Search Problem"

This story does point to one difficulty remote viewing has in both police and intelligence work—something we call the "search problem." Remote viewers can *find* missing people and things, they just can't *locate* them very well. There is a difference. Before the raid into Pakistan that lead to the death of Usama bin Laden, I was often asked during media interviews, "Why hasn't a remote viewer found him?"

"What makes you think a remote viewer *hasn't* found him?" I usually replied. Maybe that sounds a little flip, but it really is on the money. After all, it is within the realm of possibility that a remote viewer working alone might have actually located bin Laden. But that would have been the easy part. The real problem would have been trying to persuade the people who could do something with the information that the remote viewing data was worth checking out. In my experience, unless the authorities approach the remote viewer first to request his or her services, it is nearly impossible to get them to take seriously the information a viewer might have to offer.

But there is a bigger problem here, and it has to do with the nature of remote viewing. Remote viewing is *descriptive*, but not analytic. By "descriptive," I mean that what remote viewing does best is give the viewer impressions about how the target location looks: relationships between objects, shapes and activities; sensory impressions such as colors, sounds, smells and tastes; and, to some degree, spatial positioning. On the other hand, "analytic" means capable of deciphering numbers, letters, words, and so on, all of which depend solely on agreed-upon meanings assigned them by humans. In other words, remote viewing almost never produces names, addresses, latitude and longitude figures, phone numbers or any similar data that we normally use to locate things.[3]

---

3. This problem is actually a shortcoming of human psychology and neurophysiology. Remote viewing research shows that much of the experiential data is processed in modes typical of right brain hemisphere functioning, while numbers, names, addresses, etc. are governed by the left hemisphere (which, by the way, is also the source of most of the "mental noise" that adds to the remote viewing signal to noise ratio). More than 40 years of remote viewing science has shown that any names or numbers produced during the process almost always come from incorrect left hemisphere guesswork and not from more accurate right hemisphere signal detection.

In the bin Laden case, I could imagine a story that might go something like this: A viewer is asked to do a remote viewing session, fully blind to what the target might be. In the course of the session he or she says something along these lines: "I get the impression of someone with a beard and robes sitting in an odd kind of room. This room is very long, stretching out a long ways in both directions. But it's also very narrow—only a few feet wide, and with rough, rocky walls. If I go outside the room, I find myself in a dry, rocky, mountainous area with few people around."[4]

"Yes, that is exactly right!" says the analyst excitedly. " Now where *is* this bearded man?" "I just *told* you where he is," replies the annoyed viewer. "He's in a long, narrow rock-walled room in the middle of some mountains!" Essentially, the viewer has *found* Usama bin Laden but has not *located* him.

When someone hears that you claim to be a psychic (or in this case, a remote viewer), they almost always assume you to are able to do one or the other of two things. They want you to either predict the future, or to find something that's missing—or sometimes both. Unfortunately, these are the two hardest things for ESP and remote viewing to do. I will explain why that is so for predicting the future in the next chapter. But now you may be starting to understand why the second thing, locating something or someone that is missing, is also hard to do. This difficulty in locating is what we remote viewers in the military program called the "search problem."

The fact that there is this "search problem" does not mean that remote viewing is useless for finding missing things—far from it. Done correctly by someone who understands the limitations and how to get around them, remote viewing can still be helpful in locating things. Data can often be produced that are clear enough for someone to recognize the location being described and solve the crime or rescue the missing person. Could the sheriff's deputies have located the missing body in the King County, Washington case using our remote viewing data? Thanks to the hikers who fortuitously stumbled upon the body, we will never know. But the deputies themselves thought they had a good chance of doing so.

4. This obviously assumes for the sake of the story that bin Laden was in a cave or tunnel when viewed. In the interest of understanding and brevity, I have greatly simplified how this information would be presented.

Remote viewer Angela Thompson Smith, whose remote viewing work contributed to a murderer's capture and conviction. (Courtesy Angela Thompson Smith)

Rules For Forensic Remote Viewing

If you have ambitions of trying to do forensic remote viewing, you will benefit from a few points a remote viewer should keep in mind when working with the police:

The Importance of "Buy-in": You need to get official buy-in, or you might as well not even try. By official buy-in, I mean that you need to either have been already asked in advance by the authorities to remote view the case, or you need to already have a contact among them who is willing to accept your work and give it an honest look in comparison to all the other evidence that is known.

Too often, remote viewers and remote viewer wannabes think that all that is necessary is to do a session, then pass the results on to the detectives who will be ever so grateful for the inspired input. But that is not the way it works in the real world. Particularly in high-profile cases, the police are often swamped by unsolicited leads and tips passed along by large numbers of amateur psychics and remote viewers. Often this supposedly "psychic" information is contradictory or is obviously irrelevant to the case. As far as the police are concerned, there is so much "noise" from all the psychics that they can't possibly sort through it all to see what might have validity and what doesn't.

My friend, former boss, and remote viewing colleague Bill Ray worked as a civilian counterintelligence agent after retiring from the Army. Part of his job was interfacing with the senior FBI agents in New Mexico, the state in which he was assigned to work. He still remembers the day when the FBI supervisor for the state, not knowing that Bill himself was a remote viewer, went on a tirade about the huge amount of misinformation from self described remote viewers that flooded her office after every major case that came to public attention. I'm sure the erstwhile remote viewers she was complaining about just wanted to be helpful, but they should have done their homework first.

Maj. William G. "Bill" Ray (US Army, ret.), a remote viewer and former commander of the Ft. Meade remote viewing program.

As you can imagine, unsolicited input from remote viewers or other psychics usually ends up in the wastebasket. In the cases that you may have heard of where there was a successful collaboration between the police and one or more remote viewers or psychics, there were usually bridges built beforehand.

This does not mean that you shouldn't try working such cases, once you've honed your skills sufficiently. But you should develop experience and a track record first. You can do this by remote viewing the case, but recording and keeping the results only for yourself and, when the case is finally resolved, comparing what came out in the investigation to what you perceived in advance.[5] If you keep careful records—as you should—you can assemble a portfolio that will help you build those bridges with the authorities in the future.

You're probably saying, "But what if I get data that could solve the case if I could just get it to the right people?" First, you may think your data is pivotal to the case. But often, especially with beginners, what seems to be crucial information turns out to be irrelevant or even completely wrong. Establish your track record first. Learn the pitfalls of the business and their workarounds. Through training and practice learn to recognize the indicators in your own work that point to data that is wrong and data that is more likely to be right. Improve your accuracy and then approach the authorities. Trust me, another crime to solve will always come along, even if you don't get to weigh in on the current one.

Okay, let's say you now have built up a reasonably well-documented track record and feel you are ready. What do you do next? Experienced forensic psychic Pam Coronado suggests that to get your foot in the door you present your portfolio to a cold case detective. These investigators are often a little more willing than those working hot new cases to try a few unconventional leads, just in case one or another might work out. If you find a detective willing to take a chance on you, you will also have the advantage of working a case that isn't splattered all over the media, and one in which every wannabe psychic in the country isn't flooding the detective's desk with conflicting leads. You might even be the only psychic giving input on the case.

Three Important Principles:

While you are grooming yourself to assist in police work you should keep a few principles in mind. First, always pay strict heed to the elements necessary for successful remote viewing. Among these elements are "protocols" that I have already talked about—for example, blinding to the target. But proper record keeping,

5. Since you should only be remote viewing targets to which you are "blind," this must be done carefully. One way to do this would be to get a friend to assist you who will give you a variety of blind targets, while occasionally including a police-case target when you might not expect it.

honest self-evaluation of your results, and proper application of a legitimate remote viewing method are also essential.[6]

A second important principle is that without specific data, any psychic or remote viewing impression is useless, even if it is generally correct. I often receive phone calls or emails from friends, acquaintances, and even people I've never met. They tell me they have had some premonition or did a remote viewing session which gave them the impression that (as one example) terrorists are going to cause an explosion in a building somewhere in the world sometime in the future. While this may well be an accurate piece of information, it is basically useless. Most of the important elements necessary to any kind of information reporting are missing who, where, when. Only the "what" is known—an explosion in a building. Both the world and the future are huge places, and without something more precise it is impossible to take any meaningful action.

So if you can't report necessary details, you might as well keep your impressions to yourself. At least you can gain some satisfaction knowing you woro right if terrorists do blow up a building somewhere in the future. You should also take comfort in knowing that, given the limited information your psychic faculties produced, nothing you could have done would have prevented the attack. There is no need to feel guilty that you didn't pass your impressions along.

The third principle is, don't make claims you can't support. Many beginning remote viewers and, unfortunately, some of the supposed experts seem unable to resist the temptation to oversell what they can do. When interacting with others they often fail to acknowledge the limitations and shortcomings not only of the remote viewing process but of their own abilities. They raise expectations that cannot be met and, when they fail, observers come away disillusioned and skeptical. This only hurts the reputation of the remote viewer and of remote viewing itself, since critics find these kind of oversold failures useful in condemning the entire remote viewing field.

---

6. By "legitimate," I mean a method of doing remote viewing that has been shown to be successful under proper remote viewing protocols and conditions.

Making Money

What are some other uses of remote viewing? It has probably already occurred to you that remote viewing might be a valuable investment tool. I'm saving for later the story about the college students who, in just two weeks, more than doubled the money invested using an approach called associative remote viewing (ARV). I'll also mention Greg Kolodziejzyk, who between 1998 and 2009 made $150,000 using ARV in the stock market. These are essentially replications of the earliest attempts to use remote viewing to make money, when both Hal Puthoff and Russell Targ independently used ARV in separate projects to predict the silver futures market, netting significant profits. There are others who have also done well.

I myself participated in two ARV sports-betting projects, and one involving stocks that all netted profits (though the amounts were relatively small, as we had little capital to invest; when online sports betting became illegal, our efforts had to cease). More details about these three experiments come in Chapter 12.

Associative remote viewing is not the only way remote viewing can be applied in investing. A colleague of mine, Angela Thompson Smith, was for nine years retained by a wealthy East Coast businessman as an investment advisor and remote viewing consultant, until he retired from his company at the end of 2011.

He clearly felt that Angela's remote viewing contributions were worth the money he paid to keep her on board. And, though he is generally closed-mouthed about his clients, Joe McMoneagle has also provided remote-viewing based consultation that corporations and individuals apparently find useful in making investment decisions—at any rate, according to Joe, they keep coming back for more.

Digging Into the Past

One of the oldest uses for remote viewing is digging into the past, but a little differently from what the phrase might first suggest. For decades now, researchers have successfully used remote viewing as a tool in archaeology. Notable among these is Stephan Schwartz, whose book *The Alexandria Project* details his explorations using Ingo Swann, Hella Hammid, and other remote viewers to ferret out the secrets of the ancient city of Alexandria, Egypt.

Using talented remote viewers such as Michael Crichton and Judith Orloff (well known for work in their own respective fields of writing and psychiatry), Schwartz has also organized successful projects seeking out other lost civilizations, as well as undersea archaeology projects which have led to previously unlocated shipwrecks.

George McMullen, one of the viewers who worked with Schwartz, also found success with other exploratory groups. For example, in 1995, McMullen used remote viewing to find the perimeter of a stockade that probably belonged to the so-called "Lost Colony," the English settlement founded by Sir Walter Raleigh which went missing sometime between 1587 and 1590. In 2009, the "Time Team America," a production sponsored by the Public Broadcasting System, excavated the location McMullen had pointed to, and found preliminary evidence for exactly what he had predicted to be there.[7]

Archeological remote viewer George McMullen (seated) with Stephan Schwartz (l) and Ingo Swann.

7. Details about what George McMullen described and what the archaeological dig found are available at http://www.lostcolonyfoundation.org/

Emerging Remote Viewing Applications

Several practical new uses for remote viewing are in the early stages of development. Many people are exploring medical applications, particularly in medical forensics or as an aid to medical diagnosis. This is in its infancy, so we can look forward to new, interesting details as the field develops.

Others are looking into how remote viewing might aid in genealogical research. And still others are already starting to use remote viewing as an additional tool for basic science research. An example here is a recent experiment by a team looking into the properties of bacteriophages, which are viruses that attack harmful bacteria. More than 30 remote viewers were involved in the research project, and preliminary results look promising.

You don't have to be a medical doctor, a genealogist, or a biologist to participate in this work. You don't need to be a scientist or a specialist to accurately report and sketch your impressions. Each new effort generates increasing need for trained, experienced remote viewers. Though full employment opportunities in remote viewing are still rare, the remote viewing community is seeing more frequent calls for viewers to participate in these and other projects.

When it comes to applying remote viewing, the sky just might be the limit. In fact, in 1973, one of the earliest tests of remote viewing aimed to discover facts about the planet Jupiter in advance of the Pioneer 10 and 11 space probe flybys. In an experiment for the SRI research program, Ingo Swann and another noted consciousness researcher named Harold Sherman used their extrasensory faculties to collect data about the planet. In separate tries they generated a number of unanticipated perceptions, several of which were later confirmed by the NASA space probes. As one example, Ingo Swann discovered that Jupiter, like Saturn was surrounded by rings. This fact was unknown to science, yet a subsequent flyby by Voyager I in 1979 confirmed that Swann was correct.

Remote viewing has been available to the general public only since December 1995. It still counts as a frontier field and, as is usual in new frontiers, we still know relatively little about what applications and benefits a more mature understanding of the field will bring us. But that also means that much remains to be discovered, and that anyone who is determined, disciplined, and conscientious can be an explorer. Remote viewing is one of the few areas left that promise that kind of excitement to anyone who wants to give it an honest try.

Chapter 6
A Matter of Time...

In the previous chapter I said that anyone who claims to be psychic will inevitably be asked to do one of two things—either predict the future, or find something or someone that is missing. If you read that chapter you now know the pros and cons of using remote viewing to find missing things or people. So now, let's look "into the future."

Can remote viewing really predict the future? The answer to that question partly depends on what you mean by "predict the future."

"How will the world end?"

"How will I die?"

"When and where will the next major terrorist attack occur?"

If you expect remote viewing to answer these or similar questions, you may be in for a disappointment. Those are questions about the "open future" of multiple possible outcomes triggered by actions taken in the present. If you smoke that first cigarette, your cause and time of death might be different than if you don't. If you chose to stay in high school, your employment future will be entirely different than if you drop out in your junior year instead.

Despite what you hear from store-counter tabloids, the jury is still out as to whether an "open future" event can be foretold with any degree of reliability. In fact, pretty good evidence exists that predicting events such as these is largely a futile enterprise, one more of imagination rather than of certainty.

Why do people persist in linking "predicting the future" with "being psychic" when there is so much evidence to the contrary? Maybe it is because being psychic has been linked to foretelling the future almost since before the dawn of history. The Oracle of Delphi are but one widely known example of ancient seers who were rumored to predict the future. But perhaps it is because many psychics themselves claim to be able to predict the future, and many of the hopeful believe them. If you look at the data though, you discover that predicting the future is one of the hardest things for a psychic to do. Yes, some have occasional success at it, but the vast majority of attempts fail.

According to legend, the Oracle of Delphi (shown here in this 1891 painting by John Collier) was able to predict the future using her supernatural powers.

So why does the public in general still think being psychic and telling fortunes go together? Maybe it's because we tend only to remember the anecdotes of spectacular successes, while the much larger total of failed predictions gets forgotten. Some psychics themselves encourage this remembering on one hand and forgetting on the other because failure is bad for business. Remote viewing, as a so called "psychic" discipline, has to deal with the same problem, and some publicly well known remote viewers suffer from the same inclination to make predictions, then cover up their failures. Still, despite the shortcomings, there are ways remote viewing can predict certain future events with some confidence. I will get to one of these shortly.

People rightly ask how effective remote viewing is. There is for now no perfect way to answer that question, since the methods science has available can only give an approximate measure of accuracy. But a ballpark estimate based on my own research and experience says that about 70 to 80 percent of the time a well-trained, well practiced remote viewer will produce good evidence that he or she successfully accessed and described elements of an intended target. In roughly half of that 70 to 80 percent (or about 35-40 percent of the time), the results produced by a well trained, experienced viewer will not just show evidence of target contact, but will be of high quality. When the target is an "open future" event, however, these figures drop to perhaps five percent success, at best.

What does "open future" mean? Picture yourself at a restaurant, trying to decide between the shrimp and the steak. If you were to choose the shrimp, this would be the day you would discover you were allergic to shell fish. You'd end up in the hospital where you met your future spouse, the doctor who saves your life. Instead, you choose the steak and your future turns out very differently—maybe you wind up marrying your server instead. Essentially, the decision you made on what menu item to choose made one potential future outcome a reality, and canceled out the other from ever becoming real.

Similarly, in one possible future the terrorist cell plotting the next big atrocity is infiltrated by the FBI and stopped, while in another possible future the cell is not infiltrated by the FBI, not stopped, and one of the people killed in the attack would have found

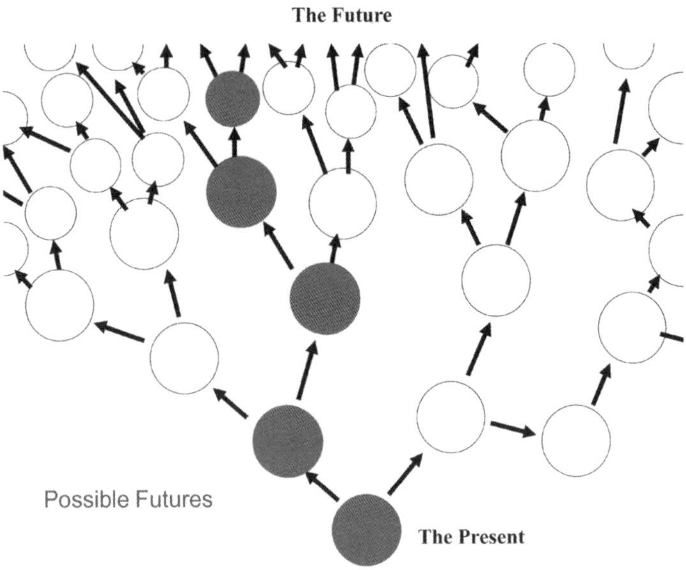

Evidence suggests that "the future" consists only of not-yet-realized possibilities. This illustration shows a simple diagram of possible futures. The circles are events that, depending on how they turn out, will decide which direction the future takes from that point. The possible future pathway (in gray) is an example of just one way the future might go.

the cure for the common cold. How a particular event unfolds in the present decides which one future gets realized, and which of uncounted other futures don't.

There are many branches of the future that are "possible." This is because events about to occur and decisions about to be made right now each almost always have multiple possible outcomes. But only one of these can actually come to pass once the event occurs or the decision is made. That means that all other possible futures attached to that event or decision will never come to be. And since which single future that will happen hasn't yet been determined before the event or decision is final, we can't know for certain which one that is and which ones will simply never occur, even if we *are* psychic. In other words, there is no actual future to "psychically predict" or remote view. It becomes a crapshoot as to what you perceive when you try to predict the future. You can see why remote viewing accuracy falls off dramatically when you try to "look" ahead.

This open array of future possibilities is what I call the "open future." Most of the events people want remote viewers to predict lie somewhere out in the vast collection of future potential events. Right now all of these events are still "open," in that for us here in the present they remain mere potentials. Eventually, all but one will be eliminated as possibilities, and a relatively small number following on from it become more likely. When someone tries to remote view an event in the "open future," it is often done under the belief either that there is such an event to be found (for example, that there might actually be a "future major terror attack") or that the manner and nature of a future event that is certain (such as when, where, and how one's death will occur) has already been decided and can be known. But the dismal record of people trying to predict these and other "open future" events shows how unpredictable such events usually are.

A few exceptions occur, of course. Some events are so certain to happen that their part of the future probably does "exist," at least as a highly-probable potential. But these are relatively rare. Because these exceptions are rare, they serve to confirm the uncertainty of most events in the potential future.[1]

Does this mean that remote viewing can never predict a future event with any success? Fortunately, it turns out that there is a way that it can. "Open future" remote viewing successes are relatively rare; at the end of this chapter I will tell you about one such rare but spectacular success that happened to me. But for now I will tell you about a way of "seeing" the future that turns out to be just as accurate as remote viewing the present or past. An added bonus is that it is easy for even novices and amateurs to do. More exciting still, this remote viewing mode can be (and often is) used to turn a profit.

---

1. Most of the cases I have in mind involve future events that include a significant amount of human intentionality, such as "the next terrorist attack." People can and do change their minds, intend things that they never ultimately carry out, or are thwarted by decisions made by still other people. Because of this, the vagaries of human intentionality make how an event turns out (or even whether it actually happens) highly variable. So any time human intentionality (free will, decision making, etc.) gets involved, uncertainty of outcome increases proportionately. (This involves a number of sometimes controversial philosophical issues that are too technical to get into here.)

"So if you're psychic, why aren't you rich?" This a favorite question frequently asked of psychics by skeptics, who for some reason seem to think they were the first to ever come up with this ancient witticism.

Occasionally, though, the question comes from someone who really wants to know. Whenever I hear the question, whether from skeptic or honest inquirer, I suspect the person asking it doesn't understand what real psychics are able to do. Since (as I explain in the previous chapter) psychic impressions in general and remote viewing in particular don't provide much in the way of names or hard numbers, it can be difficult to use the ability to make a profit. However, there is one means of leveraging remote viewing that has promise in the financial arena. I'm talking about associative remote viewing, or ARV for short.

Unlike controlled remote viewing (CRV) or extended remote viewing (ERV—both to be explained later), which are remote viewing methods, associative remote viewing is not a method as such. Instead, it is simply a strategy for tasking remote viewers against a target that won't be revealed until the future. You can use any remote viewing method to do ARV.

But there is a catch: Only certain kinds of future events are predictable by ARV. It has to be an event with a limited set of possible outcomes. Though in principle you can use ARV against an event with several such outcomes, the easiest are what I call "binary" events. A binary event has two possible outcomes, one or the other of which will certainly occur. Typical binary events include the closing price of a stock (two possible outcomes: the price will have either gone up or down in relation to the open); team sporting events (one team will win, the other lose); or the price movement of a commodity (the commodity price will either go up or down).[2]

Unfortunately, as early remote viewing researchers quickly discovered, you can't just ask a remote viewer whether IBM stock will close up or down, whether the Celtics or the Lakers will win, or whether pork belly futures will sell high or low. If you were to ask such a question of a remote viewer, the response you got would

---

2. Clever readers will note there is often a third option: for example, a stock price will stay the same or a game will end in a tie. Those are the exceptions, not the rule. There are ways of dealing with these exceptions, but for simplicity of explanation I ignore those possibilities here.

amount to little better than a guess. This is because the viewer's conscious mind ("left brain hemisphere") will weigh in, and despite the viewer's best efforts will begin speculating as to what the correct answer might be. The result will be a prediction that will rarely be any better than what would be guessed by chance.

Pioneering remote viewing scientists such as Stephan Schwartz, Hal Puthoff, Russell Targ, and Dale Graff recognized this problem. They soon realized that they could leverage remote viewing's real strength—describing a target verbally and with sketches—in a way that allowed greater likelihood of success in predicting future outcomes.

The ARV mode works like this: In advance of the event taking place, the tasker picks two objects, say a pencil and an apple, without disclosing to the remote viewer what they are. The tasker then decides that, if the stock price is higher at the close of the trading day, the remote viewer will be shown (for example) the pencil; in contrast, if the price is down, the viewer will be shown the apple. The viewer is then asked to describe what he or she will be shown after the market closes on the selected trading day. It's alright to let the viewer know the nature of the experiment—that he or she is attempting to predict up or down movement of a specific stock, or the outcome of a sporting event—as long as he or she is blind to what the specific objects were that were selected to represent winners and losers.

The viewer uses his or her preferred remote viewing method to describe the object that will be revealed several hours later, after the event's outcome is known. Let's suppose the viewer says "I see something that is round, red, sweet smelling, and reminds me of a red rubber ball." Since this sounds more like the apple than the pencil, this indicates the viewer will be handed the apple after market close. Since the apple is the signal for the stock going down in price, the tasker now knows to take action before the stock price drops. No matter what object the viewer describes, the tasker must remain true to the original plan and show him or her the object that corresponds to the actual outcome.

You can think of ARV as the viewer "looking into the future" to get some news from his or her future self—the news being the features and appearance of the object he or she will be shown after market close. By having the pencil "stand for" one outcome and the apple "stand for" the other, you have "associated" each object with one or the other of the possible future outcomes. The viewer is not being asked to predict an "open future" event that is under no one's control, such as the movement of the stock market. Instead, the viewer is seeing something that can be controlled with nearly absolute certainty, which is the act of a partner handing over an apple or a pencil.

ARV targets don't always have to be objects. They can also be photos, as long as they are different enough from each other. Here are the two target-pool photos used to predict the 2005 Superbowl the day before the game was played. The image on top stood for the New England Patriots winning.

One great thing about ARV is that you don't have to be an expert remote viewer to be successful with it. You just have to be able to remote view well enough that someone who compares your remote viewing results with the two possible target objects can determine which of them you have come closest to describing.[3] As I tell my ARV classes, associative remote viewing at its simplest is a "duet" between the viewer and the judge. If both of them are in harmony, they cannot help but win.

Associative Remote Viewing can also be used to predict events that have more than two possible outcomes. It has been used to predict horse races and state "pick-3" lotteries successfully, and people are always trying to think of new events on which to use ARV. But predicting events with more than two outcomes requires adding another target object to the set for each new possible outcome.[4] A seven-place horse race would require seven targets in the pool, for example. Thus each additional possible outcome increases the complexity of an ARV project.

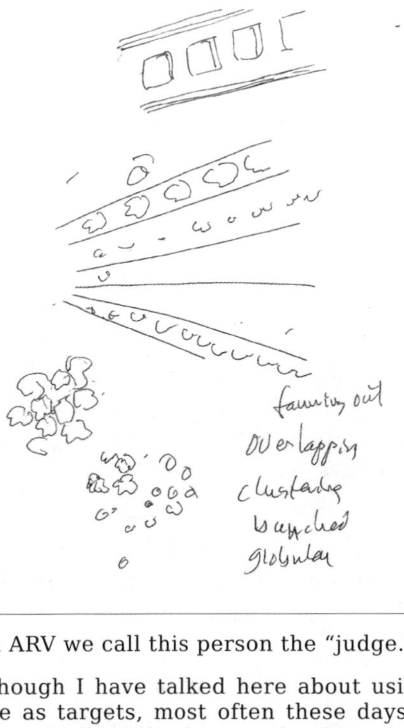

Sketch produced during the 2005 Superbowl ARV session that led to a correct prediction.

3. In ARV we call this person the "judge."

4. Though I have talked here about using objects such as a pencil or an apple as targets, most often these days ARV projects instead use photos that markedly differ from each other.

There is an active community of associative remote viewers pursuing this discipline. One of the more prolific over the past few years has been a retired physicist named Marty Rosenblatt, who has experimented with a number of different approaches to ARV. Another, Greg Kolodziejzyk, has done literally tens of thousands of ARV trials over the past 13 years, with considerable success. Still, while it is successful, ARV is no more successful than is remote viewing the present or past—about 70 percent, give or take. So what keeps ARV from being 100 percent successful?

Let's return to our previous example. What happens if the stock price goes up instead of down, yet the viewer nonetheless described the apple instead of the pencil? First, in order to keep the protocol clean, the viewer must be shown the pencil as the preselected symbol of the up market, even though he or she apparently described the apple. The standard ARV protocol requires that the viewer only be shown the associated item that matches the correct outcome, and should never even become aware of what the other object might be.

Why, then did the viewer describe the apple? The culprit in this case may be something called "displacement," which is the occasional tendency of viewers to perceive and describe the wrong associated target. Displacement occurs often enough that it can be a hindrance in some ARV projects (and lead to losing money if the project involves an investment).

Sometimes we know why displacement happens. Maybe the viewer subconsciously dislikes one target or finds the other more inviting. For example, perhaps the person choosing the targets selects a scary vampire mask to stand for one outcome and a cuddly stuffed animal to stand for the other. In that case the viewer might be subconsciously attracted to one and repelled by the other, no matter what the correct associated outcome turns out to be. Sometimes inexperienced taskers will select targets that differ dramatically in interest level from each other. An example here might be a photo of a dull, flat desert plain paired with a photo of colorful food display.

Many viewers might be subconsciously attracted to the colors and other enticements of the food image and bored by the desert scene, even if the desert photo was the correct target. Sometimes the tasker might choose objects that are too similar - for instance, where the targets selected are a donut and an automobile tire. In that case, even if the viewer's results were accurate the judge might have a hard time deciding which target had been matched. As you can see, care in choosing and matching targets is essential for ARV success.

Still, there are times when the viewer displaces to the wrong target and there is no clear explanation why. In those cases, you just have to take the loss and move on.

Other complications can get in the way of ARV success. Sometimes the judge may make a mistake in analyzing the viewer's data and choose the wrong target, even if the viewer performs well. The best way to avoid this is to make sure the judge is competent and experienced in the judging procedure.

The ARV process is very simple in concept, but can be rather complicated in execution. Nevertheless, for those who want to put the necessary work and commitment into it, ARV can be rewarding. In the previous chapter I briefly mentioned some students who more than doubled their investment using ARV. That success involved my son Christopher and his fellow classmates who, while undergraduates in a course in parapsychology at the University of Colorado, took what they had learned abut ARV and, with a $10,000 contribution in seed money, invested in the stock market. With only a bare introduction to remote viewing skills, the students predicted seven correct trades in a row and turned their seed money into $26,000. Greg Kolodziejzyk, whom I also first mentioned in Chapter 5, over a thirteen year period netted almost $150,000 in the futures market in his spare time.

In the course of a two-week period a few years ago, a student of mine named Nancy Jeane concentrated on the Texas Pick-3 lottery using ARV. During those 14 days, she got two out of three numbers in the correct order eight times, which in itself is statistically significant. But in those 14 tries she also won the full Pick-3 twice. By chance, one should win the Pick-3 only once in a thousand tries, and win twice in a row only once in a million times. The fact that she won it twice in 14 attempts is more likely than winning it twice in a row. But it still has huge odds-against chance of approximately 1 in over 71,000 tries. (Think about it this way: If you only relied on chance each time you played the Pick-3, you would have to buy 71,000 one-dollar tickets in any 14 game sequence just by guessing. In other words, pay $71,000 to earn $2,000 back. But Nancy bought a total of 14 tickets, and won the Pick-3 twice.)

Since money is on the line, it is even more important to stick close to the correct principles and protocols in using remote viewing as an investment tool. Straying from the protocol can mean you lose big money on your investment. However, when done properly ARV remains a relatively unexplored avenue for investing that has, I believe, great potential.

A word of caution here: Don't think that making a fortune with associative remote viewing is going to be easy. Doing ARV right involves careful preparation and a fair amount of work for it to be successful. It takes hours to do a competent job in a full-scale ARV project. And you have to be prepared to take the losses with the gains. In general ARV, the success rate is at best between 60 and 80 percent, even with a crew competent in ARV. And ARV is not easy to do by yourself; in most cases you will need someone to assist you. Before you risk real money in an ARV project, make sure you and any people you are working with understand the process and its idiosyncracies.

Still, the next time someone says "If you're psychic, why aren't you rich?" maybe you will be able to answer, "As a matter of fact, I am working on that right now—using associative remote viewing."

I promised you an example of when an open future predictive session actually worked. One such case took place in the Fort Meade remote viewing unit on Friday, 15 May 1987, and I was the viewer. A few minutes after I had settled into my chair in the remote viewing room in the operations building, my monitor, Ed Dames, gave me the tasking. "369147, 312200." I repeated the number verbally as I wrote it down.

My mind rapidly formed the impression of a large, grey, metallic structure moving through water. Immediately, the thought of a warship came into my head. This idea of "warship" quickly solidified into the impression of an American Navy destroyer sailing at night in the middle of a large, open body of water. The water was bounded by three shores of flat sandy land. By the time I had reached the eighth page of the transcript, I was recording perceptions of an aircraft in the distance with some relation to the ship. There was an impression of a "guttering, rumbling sound...steady" and a flying object that was long, thin and cylindrical with short, stubby wings. (The words in quotes are taken verbatim from the original official transcript.) The aircraft seemed to have "dropped and left" this object, which began to fly around a little aimlessly at first, and then headed directly towards the ship.

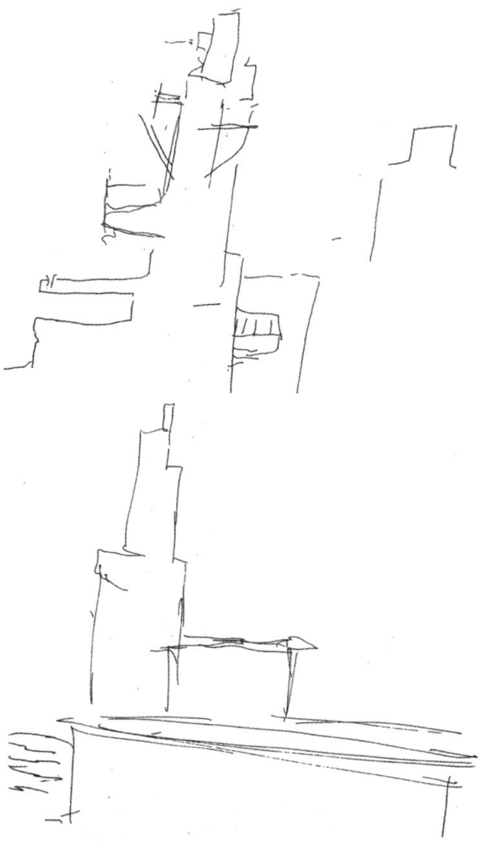

Two sketches from the May 15, 1987 remote viewing session showing the viewer's impression of a warship. Two days later, the *USS Stark* was attacked by an Iraqi fighter bomber and nearly sunk.

People on the ship became aware of the object's approach and were transfixed, as if they "can't believe this is going on." There was a collision, a "chang" kind of sound, and a "screech, metallic squeal" that "set [my] teeth on edge." The structure on the vessel "shivers, shakes, quivers," then "cants, tips, goes out of line, sags, twists." Perceptions invaded my mind of people lying on the deck, of metal, debris, fire hoses, and yelling. I had a "mental noise" impression that it was like the Exocet missile attacks on British ships during the Falklands War five years prior.

The aircraft that had dropped the flying object seemed to have come from a base in the flat, sandy land near the water, but was under control of a military in a city much farther inland, in a desert-like setting and relatively primitive by Western standards. The military personnel seemed less professional than one would expect of a Western military organization.

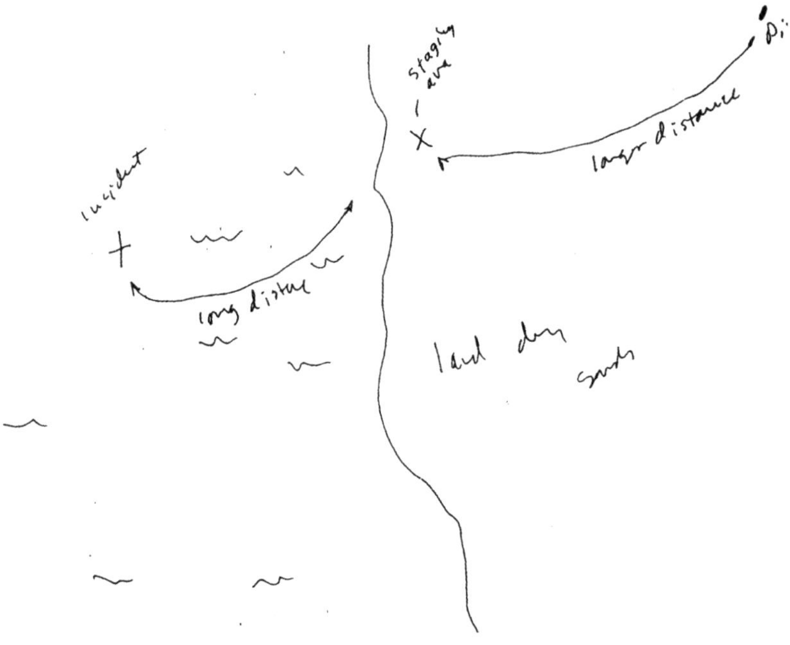

Map sketched during remote viewing session prior to the attack on the *Stark*.

Shortly after this, the monitor, Ed Dames decided it was time to quit. "I'm afraid you're off," he said. I was not reporting anything like what he had expected me to get. We were done viewing for the day. I learned later that the target number he had given me to start the session stood for something along these lines: "Describe the most important event for the United States that will take place in the next few days." So it was a tasking against a future event.

Ed had in mind a different event that he anticipated happening in the next day or two, so he thought I had "missed." Unexpectedly, though, it turned out that it was Ed who was off, and that I was surprisingly, unexpectedly, completely "on."

The work day was soon over, and it was time to go home for the weekend. Early the following Monday morning as I was getting my kids off to school, the phone rang. It was Fred Atwater. "Paul, where's the session you did on Friday?" I replied that it was in my safe drawer at the office. "Well, get in here and dig it out. We want a detailed summary of it as soon as possible!" I wanted to know what all the excitement was about. He told me to look on the front page of the morning's *Washington Post*. There in black and white was the headline: "Iraqi Missile Sets U.S. Frigate Ablaze, Causing Casualties."

The details began to pour in. A little over two days—a total of 50 hours—after I did my session, an Iraqi fighter jet had fired two Exocet missiles into the *USS Stark*, an American frigate (a destroyer-type vessel) on patrol at night in the Persian Gulf. The attack was unprovoked, but suspicions were that someone at Iraqi air force headquarters in Baghdad had made a mistake. The ship was severely damaged. It listed to one side, the superstructure was bent, tilted, and thoroughly mangled. Crew members were still fighting the fires below decks. Thirty-seven American sailors died.

A comparison with the official Navy after action report a year later showed that my session was virtually 100 percent correct in the details and description of the event.

I still shake my head at this experience. Though it was arguably the best remote viewing session I ever had, I'm almost reluctant to take credit for it, since to this day I have no real idea how it turned out so accurately. But I can't argue with the facts. Once again, the pedigree of the report is impeccable; the transcript was among those declassified by the CIA when it made the Star Gate archives widely available to the public in 2004.

The *USS Stark* on fire and adrift after the Iraqi attack. (US Navy photo.)

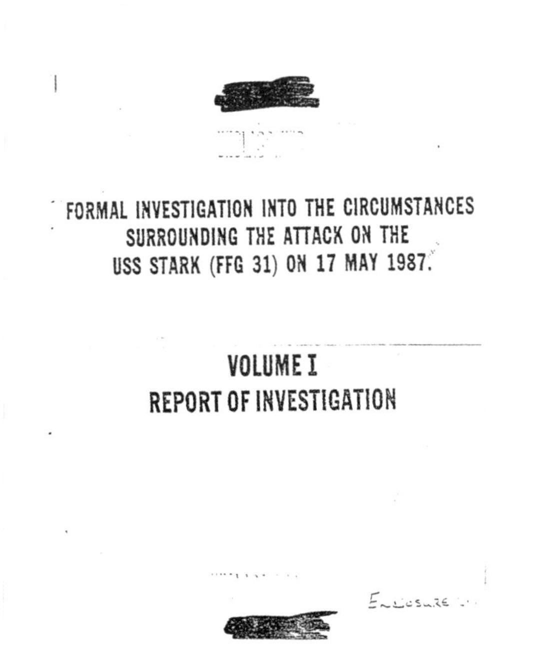

FORMAL INVESTIGATION INTO THE CIRCUMSTANCES
SURROUNDING THE ATTACK ON THE
USS STARK (FFG 31) ON 17 MAY 1987.

VOLUME I
REPORT OF INVESTIGATION

The US Navy's after-action report on the *Stark* incident.

Chapter 7
"Pick Your Poison" – the Assorted Flavors of Remote Viewing

Should you decide the world of remote viewing is interesting enough to explore further, you will find it flooded with a bewildering assortment of acronyms, nearly all including an 'RV' in the alphabet soup: There is CRV in three different varieties, along with ERV, TRV, ARV, SRV(also in more than one variety), HRVG, and so on. This chapter decodes the RV confusion for you, and in the process introduces you to how such a profusion of confusion developed.

In the beginning was GRV

Don't try to find GRV in a dictionary or online; I made it up because no other more suitable term was available. The "G" stands for "generic," and I use it as a label for any form of remote viewing that doesn't fit neatly under one or another of the more specific categories. GRV is "generic" because it has no systematic form or description in the way it is done, but instead is specific to each individual who uses it. You can think of it as "seat of the pants" remote viewing, since it is not a method that is taught, or studied, but only developed through trial and error of each remote viewer who uses his or her own approach.

I said it was "in the beginning" because this was how remote viewers learned and practiced remote viewing in the first decade of RV research and applications in the modern era. When Hal Puthoff, Ingo Swann, and Russell Targ embarked on the remote viewing adventure in 1972, viewers weren't taught how to do it. They were placed in an environment (double blind, isolation room, random target selection, no external input) where to perceive anything at all they would have to use extrasensory perception. Through a process of trial and error, they were able to develop varying levels of skill with remote viewing. Among the most noteworthy and successful of these early "GRVers" were Pat Price, Hella Hammid, Keith Harary, Gary Langford and, of course, Ingo Swann himself, who were all important contributors to the SRI International remote viewing program.

There are advocates and partisans of the various remote viewing schools who forget that some of the best remote viewing ever done was (and still is) performed under these conditions, without any formal training. To this day, no one using CRV, TRV, or any of the other RVs has ever done any better than the best results produced by the early remote viewers using the GRV approach.

Hella Hammid was involved in the SRI remote viewing program for many years, starting in the early 1970s. (Courtesy, estate of Hella Hammid, ©Tino Hammid, Los Angeles)

Pat Price (center), one of the most successful early remote viewers, with CIA official Kitt Green (L) and Dr. Hal Puthoff after a successful remote viewing experiment.

Yes it is true—as I'll explain starting in the next chapter, you can learn to remote view under the right conditions just by winging it. And you can get good at it by trying it over and over again, carefully critiquing your results each time. But it can be a difficult and tedious method, which is why CRV was developed.

Controlled Remote Viewing, or CRV

In the late 1970s and early 1980s, Dr. Hal Puthoff and Ingo Swann (both of whom I introduced in Chapter 3) put serious thought into a method of remote viewing training that would pass the skill along with a shorter learning curve. After all, you could learn to play the piano without a teacher or a method. But it is far easier and faster if you have a good method and teacher to show you the ropes. If it makes sense for piano and other skills, why not remote viewing?

Puthoff and Swann weren't starting from a vacuum. First, the SRI research showed that the vast majority of people seemed to have the underlying ability to perform psychically; they had just never tried it, nor believed they could do it. Second, enough was known about psychology to understand generally how the human subconscious interacts with the waking consciousness (what I usually refer to as one's "conscious awareness").[1] This meant that a rough theory about how extrasensory perception emerges into a viewer's waking mind could be developed. Third, by watching and documenting how successful "generic" remote viewers worked (Swann, but others as well), they had learned enough to develop a methodology for teaching other, seemingly non-psychic persons.

Great things arise out of a coming-together of smaller trends. The smaller trend in this case started when Major General Albert Stubblebine, commander of the Army's Intelligence and Security Command (INSCOM) from 1981 to 1984, wanted to come up with a way to nurture more remote viewers from among the ranks of Army intelligence personnel. Up to this point, the psychic spies in INSCOM's Grill Flame program were essentially generic remote viewers. Some informal RV methods had grown up in an ad hoc way

---

1. Obviously, not everything was known then, nor is even today about human consciousness.

over the few short years Grill Flame had existed. But growing the organization required going outside the pool of apparently naturally talented remote viewers to recruit and train viewers who might not recognize that they had any abilities at all.

To make the story shorter, I will just say that Puthoff and Swann leveraged the resources of SRI's consciousness studies lab to develop what they called "coordinate remote viewing" or CRV. I will spend more time here on CRV, since it is the direct ancestor of nearly all of the RV methodologies taught and practiced in the remote viewing community today.

CRV was first called "coordinate remote viewing" because it used geographic coordinates instead of some of the other means I'll describe later (such as beacon teams) to focus the viewer on the target.[2] From a military perspective, this made a great deal of sense. Most remote viewing targets the military would be interested in were places where intelligence personnèl couldn't get to. And if they could, you might just as well give them a camera and forget remote viewing. Years later, the term was changed to "controlled remote viewing." This is how I will refer to it from here on.[3]

CRV consists of six "stages," each progressively more involved and complex. Stage 1 launches the remote viewing session, giving the viewer a brief overall "glimpse" of the nature of the target without much detail and connecting him or her to the signal-line. "Signal-line" is the term the CRV developers used to label the stream of information that brings remote viewing impressions and data into the viewer's mind (as you may recall from Chapter 2, you shouldn't take the idea of "signal" too liter-

---

2. One or more people who go to the target as "beacons" for the viewer to focus on. (See Chapter 9.)

3. At the request of Ingo Swann, the meaning of CRV was changed in 1995 from "Coordinate Remote Viewing" to "Controlled Remote Viewing" when the military remote viewing program was declassified. Swann felt that "controlled" was more descriptive of the disciplined nature of the remote viewing approach he had helped create than was "coordinate," since that term really only described a way of tasking a remote viewer, not how remote viewing itself was done.

Hal Puthoff and Ingo Swann in Puthoff's Austin, Texas office in 2002. The two worked closely together in the late 1970s and early 1980s to develop the controlled remote viewing methodology. (Courtesy Robert Knight)

ally, as meaning anything like radio waves or electromagnetism).

Stage 1 is quickly over and merges almost seamlessly into Stage 2, which develops sensory contact with the target. In this stage the viewer perceives colors, qualities of light, smells, textures, tastes, and sounds. Soon the viewer transitions into Stage 3, which includes perceptions of height, width, shape, and volume —the target's "dimensionality," how it occupies physical space. A key feature of Stage 3 is the viewer's sketching or drawing of his or her perceptions of shapes and configurations relating to the target. Many sessions end with Stage 3, since the quality of the results may be sufficient to provide whatever information is being sought.

If more detailed information is needed, the session may progress to Stage 4, which opens up abstract or complex concepts to the viewer's awareness. For example, if the intended target is the Eiffel Tower, by the end of Stage 3 the viewer may have done a credible sketch of the criss crossing girders of the tower and identified it as a tall, metallic monument-like structure which "reminds" him or her of the Eiffel Tower. By moving into Stage 4 the viewer may begin to receive impressions of "foreign," "parklike setting," "tourist attraction," "historical," "in an urban area." In much more detailed or conceptually complex targets than the Eiffel Tower (for example, a research facility, a crime event, or a museum) several pages of conceptual data and sketches may result.

At the end of Stage 4, moving to Stage 5 is an option. Stage 5 is different from the other stages in that it is an "off signal line" mode. In other words, the viewer is not processing a real-time psychic input. Instead, this stage provides a format and procedure for digging out impressions already deposited in the viewer's subconscious by previous connection to the signal-line. In effect the viewer is "mining" a data dump of information in his or her subconscious and bringing the information out into the open. For example, let's say the viewer consciously recognizes that the target involved "tourism." This concept might itself have been based on ideas of "visitors," "sightseeing," "picture taking," "foreigners," "tour buses," and so on that came into the viewer's subconscious but—unlike "tourism" never fully emerged into awareness. Instead, what surfaced was the composite notion of "tourism." Stage 5 allows the viewer to extract the concepts on which "tourism" was based and bring them into awareness as well.

Stage 6 can often be the most impressive and interesting of all. It includes the same kind of access to complex or abstract information available in Stage 4 (as well as the more concrete, sensory-based perceptions from earlier stages), but also allows the viewer to create an actual three-dimensional model of the target. In essence, the viewer makes a sculpture of the target with modeling clay or other material. Maybe you have seen the movie *Close Encounters of the Third Kind*. In the film Richard Dreyfus, following his ESP impressions, uses mashed potatoes to build a model of Wyoming's Devil's Tower in the middle of his dinner table. This ESP-based modeling is very much like what a remote viewer does in CRV Stage 6.

The early to mid 1970s and into the 1980s was a rich era of scientific exploration into human psychology. Certain scientists were heavily involved in subliminal perception, split-brain research, insights into brain chemistry, and brain imaging. Puthoff and Swann incorporated into CRV insights from many of these fields, particularly what was then called subliminal perception and split-brain research which appeared to have direct relevance to remote view-

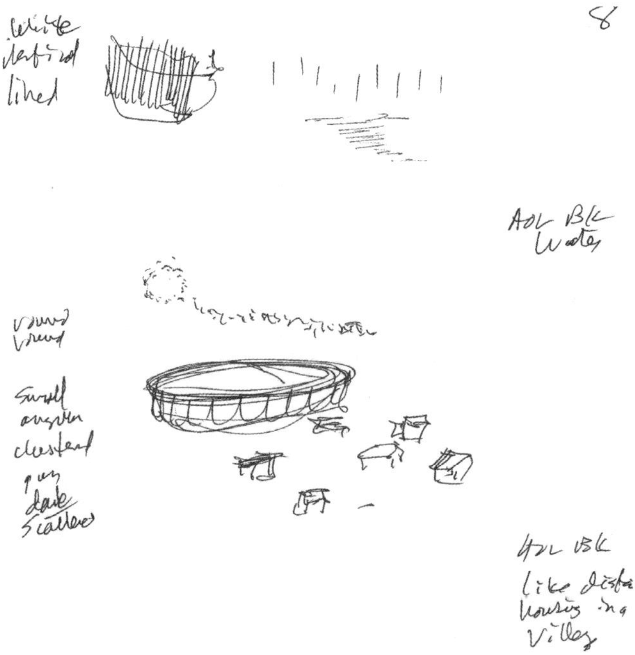

A page from a controlled remote viewing session targeted at the Beijing "Bird's Nest" stadium, compared to a photo of the target. (NASA photo)

ing. Since the point of CRV was to provide a system that, through training and practice, would help novices learn the successful behaviors of natural and "generic" remote viewers, lessons-learned from seasoned viewers were also integrated into the CRV system.

Some people complain today that controlled remote viewing seems to them to be overly complicated and rigid. But as one's understanding of the basic human perceptual system grows, it becomes clear how ingenious the CRV system really is.

After trying out the CRV methodology on a few willing human guinea pigs at SRI, Swann, acting as principal instructor, began teaching it to military personnel in 1982. Among those he taught were Capt. Tom McNear, Capt. Bill Ray, Department of the Army civilian Charlene Shufelt, and myself, also then a captain. Just as we were about to begin, Capt. Ed Dames was added to the training contract from another intelligence unit as a student. Over the next several years after our training was completed, we used our knowledge and increasing experience to train a number of new remote viewers in the CRV methodology.

Among those I helped train were Master Sgt. Mel Riley, Sgt. 1st Class Lyn Buchanan, and Capt. David Morehouse. I mention all these by name because they were destined to become familiar personalities when remote viewing later emerged from the secrecy shadows. Others we trained remain largely out of the public eye.

I'll now give brief overviews of each of the most prominent RV methodologies and their proponents. The various approaches to remote viewing can be broken down into two categories: structured and non-structured. Structured methods involve a specified format or structure the viewers follow as they navigate the remote viewing process. Non-structured approaches have little or no specific format that they follow.

I'll start with structured methods, since they are most commonly used today, beginning with CRV, together with the variations of CRV that have developed since the Star Gate remote viewing program went public in 1995.

Mel Riley, the only remote viewer to be assigned to the remote viewing program twice.

Lyn Buchanan joined the remote viewing program in April 1984, and was the first to be trained in CRV solely within the Ft. Meade remote viewing program itself.

"Structured" Remote Viewing Approaches

Controlled Remote Viewing

There are three main varieties of controlled remote viewing offered by teachers:

Problems>Solutions>Innovations (Lyn Buchanan) - Founded in the mid-1990s, P>S>I offers a form of CRV that generally follows the Swann/Puthoff format, but with modifications to several elements of the original theory and method (such as a significant alteration in the theory and practice of ideograms), as well as the addition of a number of features that were not a part of the original approach (for example, a concept called "detoxification"). The terminology used in the original Swann/Puthoff approach has also been altered. Still, most of the core CRV principles remain visible in P>S>I's methodology. Buchanan was trained in CRV by myself and others starting in 1985. As of this writing, due to a longterm family illness Buchanan's training program is in hiatus. However, some of his students have established their own training programs, offering CRV courses modeled after Buchanan's approach. Among these are Teresa Frisch and Lori Williams.

David Morehouse - David Morehouse offers a version of CRV that differs in a few elements from the original methodology, including some embellishment of the Swann/Puthoff CRV methodology. Certain details have been changed (for example, adding an extra component to the Stage 1 process), accompanied by additions of some "human potentials" mythology, but the CRV essentials are there, at least in Morehouse's published remote viewing documentation. Morehouse received his training primarily from Gabrielle Pettingell, Ed Dames, Mel Riley, and myself beginning in mid-1988. A personable, outgoing and engaging instructor, Morehouse was the first to introduce into the remote viewing story a conspiracy element, the veracity of which has since been questioned.

Remote Viewing Instructional Services, Inc. (Paul H. Smith) - Having been trained by Harold E. Puthoff and Ingo Swann in 1984, with follow-on training from F. Holmes ("Skip") Atwater and Tom

McNear, I established RVIS Inc. in early 1997 shortly after retiring from the Army. My focus has been to preserve and teach the original CRV Swann/Puthoff methodology, with only a few small updates as dictated by new scientific evidence. Two of my students, Nancy Jeane and Angela Thompson Smith offer introductory workshops and courses in CRV. Another of my students, Alexis Champion, offers courses and commercial remote viewing services in France.

Technical Remote Viewing

Technical Remote Viewing, or TRV, is an offshoot of CRV. TRV was created by Ed Dames (who received his training along with me in 1984) in the mid-1990s when he began offering training to civilians through his commercial remote viewing company Psi Tech. Due to legal problems Dames parted company with Psi Tech, but the new proprietors continued to offer TRV instruction, mostly through online and video-based means. As a method, TRV has a somewhat fluid history, evolving in the direction of simplification so that it could more easily be presented in a video training format. As TRV evolved, it increasingly modified or eliminated many of the original CRV practices and principles, perhaps in favor of making the method more

Ed Dames, who spent three years in the military remote viewing program and went on to popularize and sensationalize remote viewing for the general public. (Courtesy Robert Knight)

commercially viable. Dames still teaches remote viewing (most prominently through his "Learn Remote Viewing" video series), though using a more compact, minimalist approach suitable for weekend seminars. Well-spoken and intelligent, Dames has been a major popularizer of remote viewing, but remains somewhat notorious in the RV field due to his frequent end of the world predictions, allegedly based on remote viewing. On late night radio interview programs and over the Internet he has become widely known as "Dr. Doom," a nickname he seems to cultivate and encourage.

Scientific Remote Viewing

Scientific Remote Viewing, or SRV is an offshoot of TRV, created by Courtney Brown, an early student of Ed Dames. SRV is a more radical departure from CRV than is TRV. In addition to the early changes introduced by TRV, it also incorporates elements such as "subspace" and other metaphysical concepts that have questionable support in the remote viewing scientific literature. Nevertheless, many people have taken SRV training through Dr. Brown's Farsight institute over the years. Dr. Brown is an enthusiastic and

Courtney Brown, founder of the Farsight Institute.

Glenn Wheaton, founder of the Hawaii Remote Viewing Guild. (Courtesy Cheryle Hopton)

energetic promoter of remote viewing, but has a controversial history in the remote viewing community. He is perhaps best known outside the RV world for attempts over the years to link remote viewing to sensational phenomena such as UFOs, space aliens and the like.

Hawaii Remote Viewing Guild

The Hawaii Remote Viewing Guild, or HRVG has its own proprietary methodology that has no specific name, so it is often referred to as simply the HRVG method. The origins of this method are controversial. Some have suggested that it is a recent amalgam of various combinations of CRV, TRV, SRV, neuro-linguistic programming ("NLP") and/or the Silva Mind Method. But its proponent, retired army Sergeant First Class Glenn Wheaton, maintains that the method was developed in the early 1980s while he was in an Army intelligence assignment. As of this writing there has been no resolution to the controversy over how the HRVG method originated, but some of Wheaton's students have produced impressive results using the methodology.

Joe McMoneagle in 2011 with his wife Nancy ("Scooter") McMoneagle, director of the Monroe Institute. (Courtesy, John Alexander)

"Nonstructured" Remote Viewing Approaches

"Generic" Remote Viewing

This is a little more challenging to discuss, because by definition GRV is not something that is explicitly "taught." However, a few of those who offer guidance in workshops and seminars include remote viewing pioneers Russell Targ and Stephan Schwartz. And as of this writing the Monroe Institute in Faber, Virginia offers a "Remote Viewing Practicum," currently taught by legendary remote viewer Joe McMoneagle, that offers a general survey of remote viewing.

Extended Remote Viewing

Extended Remote Viewing, or ERV is a less widely-used method in the remote viewing field. Its origins are most easily attributed to F. Holmes "Skip" Atwater, who developed it as the method of choice in the early days of Project Grill Flame, though it saw continued use through to the end of the military program. In practice, it involves the viewer lying down and establishing a hypnaggic state, essentially "mind awake, body asleep." Once this state is achieved, an interviewer asks questions of the viewer and records the viewer's responses. Early in the Army program, HemiSync® binaural beat tapes were used to synchronize the viewer's brain hemispheres, presumably setting up an optimum state for remote viewing functioning. Later, viewers were expected to enter this state themselves.

Much less widely practiced than CRV and its derivatives, ERV is still favored by some viewers. As far as I am aware, Dave Morehouse (mentioned above) offers the only formal course in ERV.

Associative Remote Viewing

Associative Remote Viewing, or ARV, is not a method for doing remote viewing, but rather a means of tasking remote viewers to facilitate prediction of future events. As such, any of the methods described above can be used to perform ARV tasks. The previous chapter introduced the ARV approach.

I want to answer two final questions here. First, should you adopt some method or other? There is some controversy over this, with a few prominent people in the field rejecting the value of methods in general - based largely on the belief that remote viewing is not a skill that can be taught, but rather an ability one is born with. I think this mistakes the nature of the phenomenon. To be taught anything, humans must have an underlying capability. It is often said that the tricks taught to animals that perform in circuses or places like Sea World and the like are always based on a behavior the animal naturally has. The same applies to any human skill. Humans could not be taught to read if they lacked the underlying cognitive capacities and the physiological equipment (such as vision) that make reading possible. These are things humans are "born with," yet most of us (with a few exceptions) still have to be taught to read. The same applies to riding a bicycle, playing the piano, or any other complex skill humans can master. No one disputes the value of methods for helping us develop these innate capacities into usable skills.

A method's value is obvious—it gives the learner a starting point, a framework, and direction that not only guides the learning process, but also accelerates it. A well-formed method embodies the lessons learned by those who have gone before and mastered the skill, thus helping the learner avoid the usual blind alleys and false starts of the trial-and-error approach that those who don't have a method must endure. Of course, not all methods are well-constructed or properly thought out, and those methods may be less helpful, or even make things harder. I will have more to say about both these matters in a later chapter.

The second question to answer is, if you do use a method, how much does it matter which remote viewing method one uses? In some cases it clearly matters. The results speak for themselves and a simple side-by-side comparison may show either that one works better than another, or not as well-or possibly even makes no difference. But sometimes it is less clear. Often more important than method is how much dedicated practice and experience a viewer has. In such cases, is one method better than another? Probably yes, but it can be hard to tell which one and by how much.

There are a couple of reasons for this. One is because no one yet has come up with a sufficiently precise tool for measuring remote viewing success. It is usually done by comparing on a session-by-session basis the remote viewing results to the target they were meant to describe. This will tell you whether the remote viewing session was in the ballpark, and roughly how close, but it is hard to actually objectively measure the degree to which one session might be better than another. This makes it difficult in turn to compare a group of sessions from one method to those from a competing method. Useful general yardsticks exist (such as using double blind judging together with good statistical analysis to rank RV session results according to how well they match the intended target), but no sufficiently precise measuring technique yet exists.

Further, deciding which methodology is best may be vulnerable to things that aren't directly related to remote viewing at all, such as student or teacher quality, which have little to do with the intrinsic effectiveness of any particular method. In other words, one method may be better than the others, but because those who practice it do it sloppily or those who teach it are perhaps mediocre or poor instructors, the method might appear to be inferior, even when it isn't.

I will, though, share my personal opinion about methodology. As someone who has been teaching for decades, it seems to me there has been a tendency among the various practitioners and trainers to "put their brand" on the original methodology, to alter it in some way to make it uniquely theirs. A few of these alterations seem to have added some functionality; most have amounted to little more than (as it is said) changing "happy" to "glad"—in other words, cosmetic changes that don't dramatically affect the functioning of the underlying method one way or the other. But I am concerned that some changes may have significant impact on the potential for success of followers of a particular method or other. The well worn quip that one "shouldn't fix something that isn't broken" clearly applies to some of the changes and personalizations that have been made to various CRV derivatives.

Ingo Swann warned about this as he wrapped up our training in December 1984, and he reiterated it to me in a conversation in 1999. As he put it then, "Many years ago I said, 'don't change [the CRV methodology]; if it's working don't change it.' But I said everybody will come along and want to make sure that they've put their mark on it. So keeping it intact is probably going to be very hard. . .everybody wants to do it their way."

I believe that the controlled remote viewing method, as originally taught by Ingo Swann and Hal Puthoff through SRI International under government contract, is the best approach to remote viewing for teaching most people who are not otherwise extremely talented and already successful at remote viewing.

Unlike generic approaches, which require a great deal of stumbling around in the form of trial-and-error, CRV gives beginning students, many who have never had any successful ESP experience before, a path to follow from the start. As with the development of any skill, trial and error—in the form of regular practice—is still required. But there seems to be less of it necessary for the average person to acquire facility in the ability.

Since I still teach the original CRV methodology, many may see me as biased. Be that as it may, I nonetheless believe it to be superior to all the offshoots that have departed from it, whether TRV, SRV, or the assorted second and third generation versions of CRV. What I have just said will likely be challenged, and not for the first time. But if you are interested in pursuing remote viewing further, you will have to do what research you can and decide for yourself which path to try. I will give some suggestions how to do this in Chapter 15.

Chapter 8
"Can *I* Be a Remote Viewer?"

Hollywood director Elias Merhige had a credibility problem. He had just released a major feature film, with star power the likes of Sir Ben Kingsley, Aaron Eckhardt and Carrie-Anne Moss. The film had all the ingredients to make it a hit with the critics and at the box office, but it was not. It appeared that the problem with *Suspect Zero* was its central plot element—remote viewing—as practiced in the film by Kingsley playing an ex-FBI agent plagued by his personal demons. Despite remote viewing's government connections and Cold War credentials, most movie goers had never heard of remote viewing, and didn't believe it was real. They couldn't make the mental leap to take it seriously.

After his movie languished a few weeks in the theaters, Merhige knew it was too late to make it a box office smash. But as with all films, the money made in theaters is only part of the equation. Merhige had a chance to redeem *Suspect Zero* in the DVD release. He decided to add bonus features to the disk to educate viewers about remote viewing, including interviews with some of the scientists and experts from the government sponsored program. What Merhige needed for the bonus features was a credible demonstration of a successful remote viewing session. He approached me for the interview segments, and asked that I enlist Dr. Hal Puthoff, founder of the CIA's remote viewing program. Merhige wanted me to do the remote viewing demo.

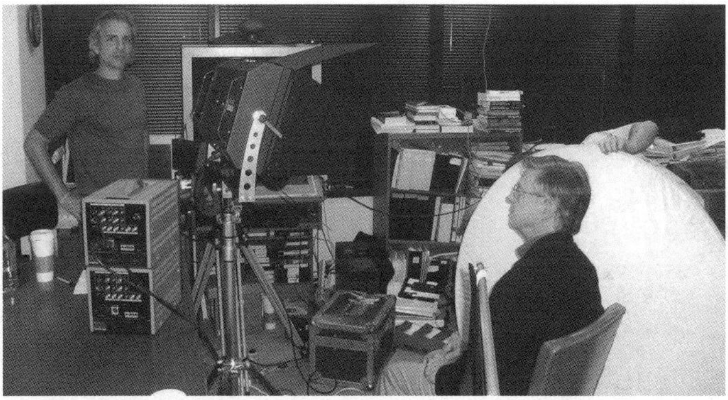

Hollywood director Elias Merhige (L) and Hal Puthoff just before the interview for the *Suspect Zero* bonus features.

I've never been a proponent of showing off remote viewing in dog and pony fashion. Demonstrating it this way isn't easy to do right, especially with a live audience, it usually isn't scientifically rigorous, and I am never happy treating remote viewing in this way almost like a parlor trick. Hal and I talked it over. We both know that even a knockout demonstration by a remote viewer is usually not enough to convince observers. They are naturally skeptical about the potential for behind-the-scenes trickery. Even those who want to believe have room for doubt.

Fortunately, in his years at SRI International Hal had a lot of experience with visiting scientists, government officials, and reporters who wanted to see for themselves whether remote viewing was real. It's the most common and logical request: "Show me how it works." Hal and his fellow researchers discovered that the best way to overcome skepticism was to make the skeptics do their own remote viewing. Only in that way could the observer be sure that he or she really was blind to the target both before and during the viewing. Those who themselves perform successful remote viewings know that it works.[1]

So when visitors to SRI asked to see remote viewing in action, Hal or one of his compatriots would say "Fine. Have a seat. Here is pen and paper." "But I can't remote view!" visitors would inevitably protest. "Sure you can. You just don't know it yet." And, thanks to what is called the "First-time Effect"—the tendency to do well the first time or two one tries remote viewing—the visitors to SRI successfully remote viewed more than 80 percent of the time.

So we pulled the same trick on Elias Merhige. When Elias walked into the room where he would interview me—where I was supposed to do a remote viewing session for the cameras—he was very solicitous. "Would you prefer to do the demonstration session before or after your interview? I wouldn't want you to get too tired." I surprised him with my answer: "It doesn't matter." "Why not?" "Because I'm not going to do the demo." I was amused by his sudden look of dismay. He'd spent quite a bit of a limited development budget to travel to my hometown of Austin, Texas, and now,

1. If you've seen the bonus features on the *Suspect Zero* DVD you may be puzzled by this. Elias Merhige interviewed Puthoff extensively. But for some as yet undisclosed reason studio editors chose to leave that footage out.

he thought, his demonstrator was flaking out on him. "I'm not going to do the demo," I repeated. "You are!" His expression, at first just disappointed, turned instantly to shock.

"I can't do this stuff!" he retorted, though he used a stronger word than "stuff." "Don't worry. Hal will help you through it. You'll do fine!" And, like most of the visitors to SRI two decades before, he did wonderfully. And you can too. In the bonus features on the DVD for *Suspect Zero* you can see the process Elias Merhige went through and the results he got that clearly astonished him. He gave a good enough description of the hidden target that as we took him to it afterwards without revealing it to him until we arrived, he recognized it on his own instantly from a distance, and was exclaiming "That's the target! That's the target!" as we approached it.

Puthoff (L) and Merhige discuss the results from Merhige's remote viewing experiment.

As this story shows, you don't have to be special or gifted to do remote viewing. And yes, you can try it yourself. It may sometimes be challenging to do it correctly without at least some limited help. (It's always easier to set up a remote viewing experiment if there is someone else to assist). But there are a couple of ways around that, which I will suggest later. One important aside: You may be interested in taking a more thorough course to learn remote viewing skills. I will discuss that option at the end of the book. For now, I will introduce you to important basic fundamentals of how to do it.

Since it will help you in the actual process, I will first describe for you what it "feels like" to do remote viewing.

What is the Remote Viewing Experience "Like"?

Have you ever run across someone at a party and had the feeling that you've met that person before? There are things about the person—appearance, mannerisms, voice—that are vaguely familiar; you just can't put your finger on how you know the person. Then, it comes back to you that you and your "new" acquaintance did in fact meet some years before.

That sense of having a "half remembered" memory is similar to the way it feels to gain a remote viewing impression. With remote viewing, though, you somehow know that it is not a memory you have from your past. A big challenge to successful remote viewing is learning to recognize exactly how that feels. Because it is called "remote viewing," people often assume that remote viewers somehow magically "see" a complete scene, or at least have clear, recognizable visual images of features at the target. However, this sort of perfect experience is more typical of Hollywood movie scripts about psychics (though, commendably, not in *Suspect Zero*). Instead, like other forms of legitimate ESP, remote viewing information comes in faintly at first, and in bits and pieces. This means that true remote viewing signals usually start off being vague, fuzzy, indistinct.

As you go through the session, you will need to notice, then record small bits of perception colors, smells, sounds, textures or tastes that you seem to perceive. Lines and shapes are also often important. Your perceptions will be fragmentary at first, but gradually, and with some practice, these impressions knit together into a more complete concept of the target as the viewer gets deeper into the session. In many sessions you may never get a full "picture" of what the target is. In fact, a fully formed, logical idea of what you think the target is will often turn out to be mostly wrong. But many of the smaller fragments will often make complete sense to you after you are done with the session.

You may wonder how remote viewing could be of any use if you, the viewer, can't make sense of what you're getting during the session. But whether you can make sense of it during the session is of little importance (in fact, some of the best and most accurate remote viewing work occurs when the data makes little sense to the viewer). After the session is done, and you can look at what you've done in completeness, it will often then become clear to you. And when doing remote viewing work operationally—that is, in support of some real world project—the data you provide may be the very pieces of the puzzle the analyst who sees your work will need to solve the problem you are trying to help with.

If all there was to remote viewing impressions was recognizing "bits of perception" and "half-remembered memories" that gradually become more distinct and organized, then it should be fairly easy, right? There is a little more to it—but that "little" turns out to be rather large. Remote viewers don't just get signal; they also get mental "noise." As the father of information theory, Claude Shannon, discussed long ago, every communications channel has "noise" mixed in that often gets in the way of the signal. Remote viewing, involving a perceptual information "channel," is no different. You should always be on the alert for noise, but with a few practice sessions you will start to discern the difference between the genuine signal and the noise. This is why it is important as you go along to make sketches of what you think you are perceiving. Quite frequently, sketches that don't appear to make much sense when you first put them on paper turn out to be fairly accurate depictions of part or all of the target.

Remote viewing impressions must compete with the mental "noise" that occupies your mind all of the time. Mental noise is made up of the memories, thoughts, worries, guesses, deductions and distractions that keep your brain buzzing. Sorting this out from the true remote viewing signal is the hardest part of the whole process. There are some helpful guidelines:

1) Watch out for mental images that are bright, sharp, clear, and static; these are almost always "noise," and therefore mistaken information.

I know this sounds counterintuitive. Why call it remote viewing if what you clearly "see" turns out to be wrong? But not everything you "see" during the remote viewing process is necessarily true. Often mental imagery is created when your conscious awareness struggles to explain perceptions that are developing deeper in your mind from the raw data the remote viewing signal provides.

I often talk about remote viewing being a "low band width" process—that is, the ESP signal is weak and subtle as it enters into your mind. Modern science has learned how the two halves of your brain interact. The left hemisphere is home to what prominent experimental psychologist Michael Gazzaniga calls the "Left Brain Interpreter." Its job is to explain to you what the bewildering variety of sensory input you receive every second means for your survival and success. And the left brain usually does its interpreter job very well. But when the sensory input signal is weak (as it is in remote viewing) and carries only small amounts of information at a time, the left brain often gets its interpretations wrong. This is why a viewer might be targeted at, say, the Eiffel Tower, with all its crisscrossing girders, yet consciously decides that the target must be a steel girder bridge. Be aware of this natural tendency of the human mental system, and don't be discouraged if it takes awhile to learn how to deal with it.

2) Let go of your need to succeed.

In my experience as a teacher of remote viewing, the students who struggle the most are hyper analytical types who believe success is a natural outcome, and they won't rest until they attain it. It is no accident that introverted electrical engineers often seem to have the toughest time.[3] But no matter who you are, to succeed at remote viewing you must be willing to fail. This too sounds counterintuitive, even contradictory. But it is absolutely true. If you are afraid that you are going to fail, you will hold on too tightly to the process, become too invested in the outcome and, as they say in the sports world, "choke." You have to be willing to relax; take the failures with the successes as a matter of course. Try things, take intuitive risks, trust impressions, acknowledge a fleeting thought that doesn't "make sense" to your conscious mind. Trial and error is the only way to gain proficiency at tell-

3. Fortunately, those who follow the remote viewing rules of the road (and, of course, practice) almost always have success eventually.

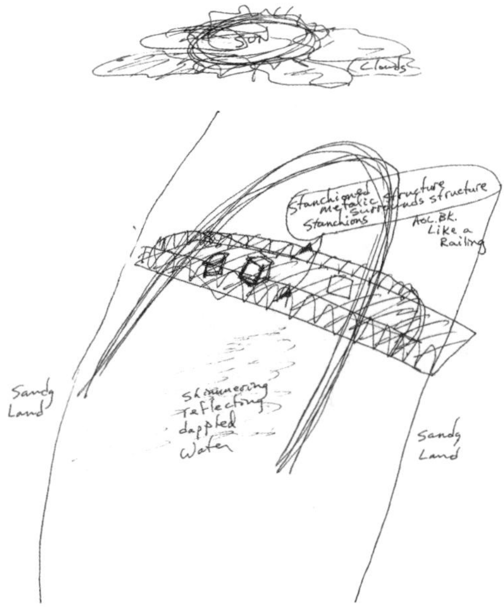

A student's remote viewing session, with the Hohenzollern Bridge in Cologne, Germany as the target (below). When looked at in isolation, the sketch looks confusing and imprecise. When compared to the actual target, it easy to see the family resemblance between remote viewing impressions and the target that inspired them.

(DKuru/Wikimedia Commons)

An image in your mind that is clear, sharp, bright—a "picture postcard" sort of image, such as this one—is usually generated by mental noise and is almost always wrong.

ing the difference between correct and incorrect data. Once you reach the point where an unsuccessful remote viewing experience is just as interesting and educational for you as a successful one, you will find your successes beginning to outnumber your failures.

3) Realize that remote viewing doesn't always work.

Going hand in hand with being comfortable with failure is recognizing that even the best remote viewers don't "get it right" every time. The successes will come more often with practice. Still, you must be able to recognize and admit when things go off track. Far too often I see beginning viewers who are seemingly unable to admit that remote viewing didn't work for them in a given instance. They try to make it appear as if they had a hit instead of a miss. I call this "data fitting," when a viewer tries to shoehorn poor remote viewing results into matching the target, even when there is no real match to be found. Skeptics call this "confirmation bias," the tendency to see what you want to see, even when everyone else can see that you are wrong.

This attitude will cause trouble for you in a number of ways. It will give your skeptical friends reason to roll their eyes while questioning your objectivity and critical thinking skills. But much more important than that, it will hold you back in your remote viewing progress. If you find it difficult to admit the difference between whether you've made a hit or a miss, you won't improve and your data will not be very useful. Learn to recognize when your data is wrong, and own up to it—don't try to pretend that you've succeeded when you haven't.

4) Keep complete records.

Finally, keep good records so you can monitor your progress. You should always keep your session transcripts together with the feedback photo of the target that goes with it. Always date and put your name on everything you do, then file it in an accessible place. Why is this important? Because this is the only way you can measure your progress in the longterm. If you don't measure your progress, you won't know if you have improved. And record keeping has an added benefit. When inevitably a friend or relative questions your sanity for being involved in such an obviously impossible waste of time, you will have the evidence you need to show them they were wrong. You'll be surprised how satisfying that can be!

The bottom line? Success depends on two things:

1) Learning to recognize the signal; and

2) Learning to manage mental noise. There are many other principles and details to remote viewing, but this is sufficient to get you started.

Keeping a continuous record of your remote viewing attempts is important for improving your performance and tracking your progress.

Enough detours! How do I remote view?

There are some who claim they can teach you everything you need to know in a few minutes or hours. They are wrong. To be sure, you can learn how to do a simple remote viewing in a reasonably short amount of time. That's what I am going to teach you here. But I will be honest. I can't teach you everything about it in this short chapter, or even in a complete book. No one can. Much of what you need to know can only be taught through the experience of doing many remote viewing sessions yourself. In addition to practice, there is much theory and procedure to learn so you become fully informed and competent in the remote viewing process.

Indeed, there are courses you can take to help you develop sophisticated remote viewing skills more quickly than you would be able to develop through trial and error (how you choose from and evaluate these courses is what Chapter 15 is all about). Right now, I will give you just enough to get you started and provide a taste of what remote viewing is like. I think you will find that is plenty to bite off at this stage.

Your remote viewing session starts with a cue—a sort of "ready-set-go" that alerts your consciousness to begin remote viewing. If you have someone assisting you (this person is usually referred to as a "monitor," or sometimes "interviewer"), he or she may give you the cue. If you are working alone, you will have to provide it yourself. In most simple cases such as you will learn here, this cue may be nothing more than the word "target" spoken aloud and written down in the left margin of the paper.

At this point, simply let your mind drift to the target. Since you will be working under blind conditions, you will have no direct conscious awareness of what the target is. Open your mind to what might be present at the target in terms of the most basic sensory impressions. Resist the temptation to speculate about what the target might be. (Any time you find yourself wondering, "What is it?"you're inviting your analytic mind to make something up.) Instead, pay attention to what impressions of color, smell, sound, texture, taste, or shape come in to your awareness. Describe those sensations on your paper. Sketch the shapes and lines you have impressions of. If you get any specific mental images or impressions of "what the target is," jot them down on the right side of the page.

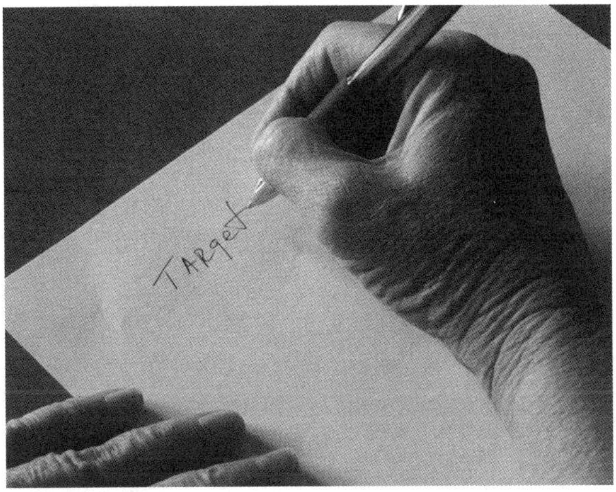

For simple remote viewing experiments, merely writing the word "target" down as a mental starting cue may be enough to get your session going.

This is where you record data that comes through "mental noise." You will be able to sort these out later to see if they contain any useful impressions.

Some remote viewing methods have specific procedures for where on the page to record categories of impressions, but those are not important for a simple remote viewing such as this. Just keep things in some order, progressing down the page (and onto additional pages as necessary) as more impressions come in.

Earlier I described remote viewing impressions as being like faded memories—vague, yet with a certain real quality about them. Novice remote viewers often find it difficult to recognize these impressions, expecting something more definite or clear. Here is a rule for beginners: Record everything on your paper that comes into your mind that might have anything at all to do with the target. If you get an impression of the color red or a smokey smell, write it down. If you sense a square shape, draw a square. Elaborate on your sketch, or even make new sketches as you perceive more details. It doesn't matter if you have no artistic ability. Sketches are important conveyors of information whether or not they have any "artistic" merit. Even if you are unsure of what you are experiencing, jot down the impression. It is often the case that the very things a remote viewer doubts turn out to be accurate.

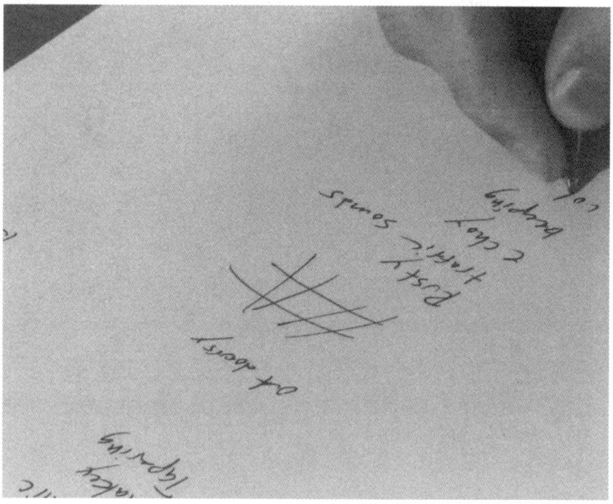

It's helpful to record anything you think may be mental noise (and not real impressions from the target) in the right hand margin of the paper. This separates out the noise and allows you to go back after the end of the session to evaluate how close or far from being related to the target the noise was.

A remote viewer sketching his impressions of a target during a remote viewing session.

You were probably expecting to get something more impressive than "red" or "square." You will. But even seasoned remote viewers usually start every session at this basic level and then progress through the session to more sophisticated information. After a few pages of words and sketches, you will begin to develop a general mental concept of the physical nature of the target. This you should sketch with as much detail as you can without your left brain making things up. Be patient. You will eventually learn to recognize when the analytic side of your brain is trying to be too helpful by jumping to conclusions on your behalf. This recognition will be of great help in sorting out the real data from the false.

When you feel that you have gotten as much information about the target and made as many sketches of it as you can, simply write the word "end," along with the time you finished. (If you feel you have enough information to warrant it, you may choose to write a short summary of what you received.)

After you are finished with the session and the target has finally been revealed to you, go back over your session and compare your impressions to the truth about the target. Mark those things that you got right, the things that were wrong, and the things that you are uncertain about. For example, let's say you got impressions that the target was red, black, rough, tall, metallic, cold to the touch, and rounded like a cylinder. Then after you were finished, you learned that the target was rectangular, not round, and that there was no red there. But your descriptions of black, rough, metallic, tall, and cold were accurate. One way you can keep track of what is right and wrong in the session is to circle what you got right and draw a single line through what you got wrong. (No erasing or blotting out allowed!)

Now, think back to when those impressions—correct and incorrect—arose in your mind, and try to recognize the subtle differences in how each felt as it emerged into your conscious awareness. You will find that as you go from one remote viewing experience to the next, though there may be additional setbacks, the percentage of correct impressions will increase and the incorrect ones will become fewer (though they are likely never to go away entirely).

What about all those images or full blown impressions of "what the target is" that you have written on the right side of your session transcript? These constitute mental noise, or "analytical overlay"

(AOL for short). You will quickly realize in the post mortem on your session that these AOLs are almost always wrong—especially early in the session. But you may also discover that AOLs you noted later in the session have a much stronger resemblance or relation to the real target. This can be helpful in learning over time to recognize when you're "on target" with your impressions and when you're not.

Now you are ready to take the plunge.

CASE

Chapter 9
Two Simple Remote Viewing Experiments You Can Do

If you have a desire to learn remote viewing, you should find the words of two of the earliest remote viewing pioneers encouraging. As Dr. Harold Puthoff and Mr. Russell Targ explicitly state in their path-breaking book *Mind Reach: Scientists Look at Psychic Abilities*:

> In our experience, anyone who decides for himself that it is safe to experience paranormal functioning can learn to do so. In our experiments, we have never found anyone who could not learn to perceive scenes, including buildings, roads, and people, even those at great distances and blocked from ordinary perception. (Targ & Puthoff, *Mind Reach*, p.5)

With your knowledge of the basics, it's time for you to try remote viewing. In this chapter I will outline how to do two different experiments. One you can do by yourself with just a minimum of help up front. The other takes a bit more assistance from friends or family, but otherwise is easy to accomplish.

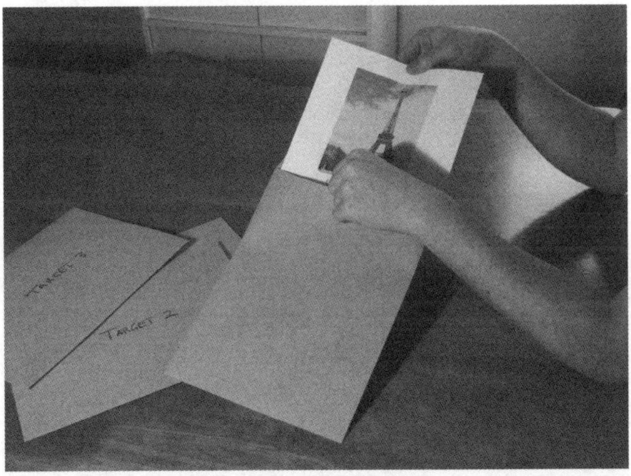

Getting the target pool ready by sealing different target photos inside identical opaque envelopes.

Sealed Envelope Remote Viewing Experiment

One simple approach is to remote view a photo that is hidden in an opaque, sealed envelope. The first task in setting up this remote viewing experiment comes in arranging for a target. As a viewer, you can't choose this target, or know anything about it until after the remote viewing session is over. This problem of target selection exists for any approach to remote viewing.

Have a friend or acquaintance select several interesting photos with strong shapes, lines, and colors. Your friend should then glue the photos on plain sheets of white paper, and seal each photo in a separate opaque envelope; it is important that nothing of the contents shows through to the outside and all of the envelopes must be identical in size and appearance.

Your friend should number the envelopes sequentially from "1" to whatever the highest number of target envelopes he or she has created. You, yourself, should have no idea which numbers correspond to which photo, or even what the photos are. Remember, this is what we call being "blind" to the target.

The photos should not be too complex, but striking enough that they will hold some interest to the remote viewer's subconscious mind (which is heavily involved in the process). It is also helpful if the photos are as different as possible from each other. The beginning viewer's results may only partially describe elements of the photo, so you want to make it as easy as possible to distinguish which photo the viewer has described when the session is over.

This would be a poor choice for a remote viewing target. There is no real center of interest and overall it is too uniform to be attractive to a viewer's subconscious.

This, on the other hand, would make an excellent remote viewing target. There is strong contrast between line, shape, and color, and the subject itself is attention-getting.

When you are ready to do the session, select one of the envelopes at random, and sit at a table in a quiet place. You will need several sheets of paper (standard white office paper works fine) and a black ink pen. Jot down your name, present location, the date and time in the upper right corner of your paper. Begin your session by writing "Target 1" (or whichever envelope number you have selected) at the top of your paper. That will be your "ready-set-go" signal.

Relax and try to perceive the impressions that come into your mind from the photo in the envelope. You should write down impressions such as colors, shapes, textures and smells that you can capture with words. Other impressions that are not so easily described verbally should be sketched or diagrammed on the page. Use as many pages of paper as necessary to give a complete account of what you perceive during the session. Some viewers find that in their first few attempts they have difficulty noticing any perceptions, while others get a flood of information seemingly without trying. If you find you're having difficulty receiving or recognizing impressions at first, simply keep trying. Each time you do a remote viewing session, relax and explore your mind for any subtle impressions you might not have noticed previously. The fact is, the impressions are almost always there, but some people have never noticed them or learned to recognize them before. You can develop that skill.

When you are ready to start your experiment, use a die to randomly select which target envelope you will remote view.

Ready to start!

A novice remote viewer doing the experiment.

Something else that is crucial to understand: since the human perceptual system tends to drift away or be distracted from things the person is trying to concentrate on, it is always important every few minutes or even more often to reinforce your attention on "Target 1" (or whatever target you are working on at the time) while you are trying to remote view it. You can do this by simply emphasizing "Target 1" in your mind, or even quietly verbalizing "Target 1" to yourself when you notice your attention drifting. Always drawing or sketching what you perceive is just as important. I can't emphasize enough the value of sketching, as it is a way to bring out impressions that develop in your mind that may be hard or impossible to explain with words. In fact, sketches often turn out to be more significant than words in expressing what you discover.

This sort of experiment should only take about five to ten minutes. When you feel you have gotten all you are going to get from your target, write "End" and record your ending time at the bottom of the last sheet of paper you have used. From this point on, you should make NO FURTHER MARKS on your written remote viewing record. (This is called the "transcript" of your remote viewing).[1]

---

1. I earlier recommended you mark right or wrong elements on your session transcript, yet here I am saying not to mark on your paper. In a learning mode, scoring marks are okay – but only so long as they are clearly scoring marks, and don't change the meaning or content of anything you have written or sketched.

When finished, you should compare your results to the intended target. Be honest give yourself credit for the things you got right; but don't try to make the things you missed "fit" the target in some way.

You may now open the envelope to see what the target was and compare it to your session transcript.

Be honest with yourself. Where something matches well, give yourself credit. But don't strain to find a correlation between what you "viewed" and the target photo. Beginning viewers are often tempted to stretch their results to match the target or portions of it, even when there really isn't a match at all. As I mentioned in the previous chapter, this is called "data fitting," or "confirmation bias," and is usually just a way of making excuses for yourself. Data fitting can get in the way of improving your remote viewing abilities since, if you can't acknowledge where you've been wrong, it's harder to learn how to do things right the next time.

For variety, this experiment can also be done by having someone hide an object in a box or other opaque container. The viewer then attempts to provide verbal and written descriptions, and sketches of whatever has been hidden in the container. (The viewer shouldn't be allowed to handle the container until the session is done.)

"Outbounder" Remote Viewing Experiment

Another easy experiment is the one that Hal Puthoff and I used on film director Elias Merhige, and has been commonly employed in everyday laboratory settings both to demonstrate and to research remote viewing. This is called an "outbounder" or "beacon" experiment, and was originated by Ingo Swann and later refined by Puthoff and Targ. In this experiment, the remote viewer will describe and sketch details of a randomly-selected physical location. The viewer "finds" the target by mentally homing in on a "beacon team" consisting of one or more people who have gone to a previously selected location about which the viewer knows nothing. Targets that have been used in past experiments of this type have included playgrounds, public buildings, boat marinas, windmills, unique natural landmarks, commemorative monuments, and so on. Just about any location with distinctive features can serve as a target.

Before setting up this experiment, let's review the basic remote viewing rules:

- The viewer should never be told what the target is or any thing about it until the session is finished.

- No one with the viewer before or during the session should know what the target is, either. Following these two rules sets up what is known as a double blind condition.

- The viewer should be placed in a situation where he or she can relax, and where outside interference (loud noise, brights lights and colors, etc.) is kept to a reasonable minimum. A quiet, comfortable living room, home office, bedroom with a desk or table, or similar setting would be appropriate for this.

To make things as clear as possible, I'm going to use a step-by-step "recipe" format to explain how an outbounder experiment works.

Overview

A remote viewer in a closed room and having no knowledge of the target, uses extrasensory perception (ESP) to perceive and describe a target location where one or more other persons (the beacon team) have gone.

Participants

- A remote viewer
- An interviewer or "monitor" (helpful, but optional)
- A "beacon" or "outbounder" team consisting of one or more persons

Materials

- Several target envelopes (four to six is a good number) and a die
- A stack of white 8.5x11 paper and a pen with black, indelible ink, medium point
- An automobile to transport the beacon team if the target is too far away to walk
- A quiet place with a table and chairs where the session can be conducted

Preparation

Someone not directly involved in the actual remote-viewing part of the experiment selects four or more possible targets (in an informal experiment like this a member of the outbounder team can select the targets, but it should not be the interviewer, and never the viewer). The possible targets should be within about 15 minutes travel time more or less from the viewer, for convenience.[2] The name, location and driving instructions to each separate target are put into that target's envelope and sealed. (It is useful, but not necessary to include a representative photo of the target.) Each of the target envelopes should be identical on the outside to all the other envelopes in the set. Finally, the envelopes must be thoroughly opaque so nothing can be seen from the outside, and there must be no identifying features on the outside of any of the envelopes.

---

2. Beacon targets can be farther away. In past experiments they have been as far as on the other side of the world from the viewer. However, it simplifies things if the target is closer rather than farther away.

Targets

The targets should be as different as possible one from another, with as few features in common as practical. It may be difficult to eliminate every common feature, but do the best you can. This is important so that by the end of the remote viewing session it will be as obvious as possible whether the viewer has described one target or another. An example target set might include a bridge, a library, an amusement park carousel, and a bakery (note that each target is significantly different from any of the others). Use your imagination, but don't pick targets that are too complicated — that is, have too many different features and details associated with them (for example, an entire city). Care should also be taken if at all possible to not select well known local landmarks that the viewer might be tempted to guess, such as the Golden Gate Bridge, the Eiffel Tower, or the Washington Monument.

Procedure

Small variations on the process are allowed, but it should proceed somewhat along the following lines:

1. The beacon team, the designated remote viewer and the interviewer gather at the place where the viewer will work. Watches are synchronized, and a time to begin and end the viewing is agreed upon.

2. The target envelopes are shuffled, then someone arbitrarily numbers them. Another party rolls a die to determine which envelope is selected. (If you have fewer than six target envelopes, keep rolling until you get a usable number.)

3. The beacon team takes the selected envelope but DOES NOT open it yet. They proceed to a location where—out of sight and hearing of the viewer and monitor—they open the envelope and follow directions to the target.

4. In the meantime, the remaining envelopes are put away where the viewer cannot have access to them. The viewer and interviewer enter the viewing room and wait for the agreed upon time.

5. The beacon team travels to the target. If they arrive before the agreed upon remote viewing session start time, they must wait to approach the target until that time arrives.

Jamie Conrad was part of the outbounder team in our experiment with Elias Merhige.

6. Once at the target, the beacon team will attempt to interact with it as much as possible. If, for example, the target is an amusement park carousel, they might look at it, ride it, stand near it, touch it, talk about it, etc. This lasts for 15 minutes, at which point the team will return to the car and drive back to the viewer's location. Keep in mind, this isn't an exercise in telepathy; the viewer is not trying to "read the minds" of the beacon team, but only to home in on their location. Their activities at the target may serve to make the location more attention-getting for the viewer.

7. At the moment the beacon team approaches the target, the remote viewing session begins. It lasts the entire 15 minutes the beacon team is at the target site. This means the viewer will verbalize, write down, and sketch any mental impressions that might have to do with the target. The interviewer assists by prompting the viewer to direct his/her attention around the target. Since the interviewer knows nothing about the target, these prompts will be general in nature. For example, the monitor may say things like "What colors or shapes do you notice?" "Listen for important sounds at the target." Or "Can you sketch what the target might look like seen from above?" and so on.

8. At the conclusion of the session and after the return of the beacon team, the remote viewer is taken back to the actual target so he/she can receive feedback as to what the target was and to compare what was perceived during the session to the actual target.

After the experiment, Elias Merhige accompanied the outbounder team to the target to compare his results and perceptions to the actual location.He was impressed by the outcome.

(Alternatively, if the return trip is impractical the beacon team can take a video camera along to the target to record the experience while they are there. The video can then be played back to the viewer for feedback purposes.) In a pinch, the viewer can visit the target alone. But having the beacon team present for the feedback can help the viewer understand what was going on at the target when the team was there during the experiment.

That's all there is to it, and it is essentially what we did with Elias Merhige. He was thrilled with the results, and very happy that I had "tricked" him into trying it himself.

[NOTE: As few as two people can set up and carry out an outbounder remote viewing exercise. But it is easier if you have three or more. This is especially true if you want to make this experiment truly scientific. In that case, you can arrange to have an additional person who has nothing otherwise to do with the remote viewing itself or otherwise uninvolved with the experiment to pick the target set. That person, or an additional person (or persons) who is also otherwise uninvolved with the remote viewing or the experimental process can then blind-judge the results. This means that a judge is selected who has had no contact with either the viewer or anyone else involved with the remote viewing experiment. This judge is then shown what all of the targets in the target set were. Without knowing which target was the one the viewer described, the judge then tries to match the correct target to the viewer's results. These kind of matchings can be statistically analyzed and presented as legitimate scientific results.]

Chapter 10
How Does Remote Viewing Work?

One of the most common question I am asked about remote viewing is, "How does it work?" The most honest answer I can give is, "Nobody knows;" or, as my friend Bill Ray always says ever since we started our training together more than three decades ago, "Magic." But inquiring minds aren't very satisfied with those answers. Fortunately, there are some speculative theories about what makes remote viewing work to point us in the direction of a possible answer.

It helps if we first eliminate what *doesn't* explain how remote viewing works. ESP (and thus, also remote viewing) involves what is called "action at a distance." In physics terms that means one thing causing an effect (a movement, a reaction, a transmission of knowledge or information) on another thing that is distant from it, without any direct cause linking the two.

According to classical physics, true "action at a distance" is supposed to be impossible. Basic physical principles can explain some of what at first appears to be "action at a distance." For example: A boy on the ground with a device in his hands guides the flight of a radio-controlled airplane with no visible connection between the two; or you and a friend have a cross-town cell phone chat with no wires connecting your phone to hers. In either case, even if no visible cause seems to link either end of the action, we know very well that this is not real "action at a distance," since radio signals, invisible to the human eye, connect one end of these actions to the other end with an unbroken causal chain.

In remote viewing, the "action at a distance" is the remote viewer's mind receiving perceptions from a distant or hidden target. The difference between the model airplane or cell phone examples and remote viewing is that the connecting causes are known in the case of the phone and the airplane; both involve radio signals. In remote viewing and other forms of ESP, we have almost no idea what the connecting causes might be.

As I explained in chapter 2, we can be sure that electromagnetic sources, such as brain waves, radio or microwave signals don't have anything to do with making remote viewing work.[1] That has

1. Some might object that this isn't altogether true. Obviously, brain functions are necessary to process the ESP signal, and the brain works on electrochemical activity. However, it is only the ESP signal that we are concerned with when we say that electromagnetism is not involved in ESP or remote viewing.

been ruled out by scientific experiments. And magnetic fields aren't the cause, since such fields can't carry the kinds of information, nor can they affect human brains the way ESP seems to. (And magnetic fields can't account for how ESP signals can transcend time.) Further, it's obvious from simple physics and biology that ESP signals can't come through any mechanical means; there is no "tin can telephone," or postal service that brings ESP signals to the perceiver. In general, that covers just about everything science knows about "action at a distance," and none of it explains what makes remote viewing "work." In laboratory studies, each of these measurable causes were eliminated from the process and remote viewing still succeeded.

There is (for now) one other possible explanation that has been suggested by parapsychologists who also understand physics. Among these are Dean Radin, Ph.D., lead scientist at the Institute of Noetic Sciences; nuclear physicist Ed May, Ph.D., who took over leadership of the SRI program in 1985; Nobel laureate physicist Brian Josephson; and of course, Dr. Hal Puthoff, who the CIA commissioned to found the original remote viewing research program and is author of one of the earliest textbooks on quantum electrodynamics. And there are others.

What they suggest is a physics phenomenon known as "quantum entanglement" or "quantum nonlocality." The term entanglement is used because it appears that (at least at the subatomic level) two particles can become "entangled" in a relationship

to the extent that, if you affect one, the sister particle is affected as well. It is often called "nonlocality" because the change in the sister particle takes place instantaneously and regardless of how close or how far apart they are from each other. More significantly, according to quantum physics there is no causal connection linking the behavior of the two particles it—"just happens."[2]

Physicists Brian Josephson (opposite page) and Dean Radin (above) are two of several scientists who think that quantum non-locality may provide an explanation for ESP and remote perception. (Josephson/Wikimedia; Radin/Jim Johnson)

Think of it this way: The basic building-block for light is a tiny particle known as a "photon." A physical process such as collision or splitting will "entangle" two or more photons. Imagine that we separate those "entangled" photons so they are a long distance apart. Even inches apart is long distance for something as tiny as a photon, but to make this particularly impressive, let's separate them by miles or even light years instead. According to another rule of quantum physics, particles don't have defined properties until they are "observed" or measured. They just exist in a state of all potentials. When we measure a property of one of these pho-

2. Both "entanglement" and "nonlocality" can be used interchangeably to refer to this bizarre physics phenomenon, but I will stick to "nonlocal" and "nonlocality" in what follows.

tons, perhaps its direction of spin, the sister photon's spin suddenly becomes measurable, too even though we've not done anything to it. This seemed so much like "magic" that Albert Einstein called the principle "spooky action at a distance."

Albert Einstein thought quantum nonlocality was too much like "spooky action at a distance." (Ferdinand Schmutzer photo)

He was uncomfortable with the whole concept because it conflicted with his ideas about how the universe was supposed to work. Yet nonlocality is a real principle, accepted by almost all physicists. In what has become known as "Bell's Theorem," physicist John Stewart Bell showed mathematically that nonlocality holds its own among the best established physics formulas, and recent well-designed physics experiments have confirmed it.

How does this relate to remote viewing? First, many people have noticed similar characteristics between both quantum nonlocality and the various forms of ESP. Both involve actions at a distance that show no connecting causes. Both seem to be instantaneous. Neither seems to diminish in effect even as distances increase between what is acting and what is being acted upon.[3]

3. We do have to be careful with words such as "acting upon." A physicist would say that in a nonlocal relationship nothing is acting on anything, since nonlocal action is just a "correlation" between effects—one event occurs, and so does the other one, every time.

John Stewart Bell proved mathematically that quantum nonlocality was real. (Queen's University Belfast/Wikimedia)

Second, the brain is driven by electrochemical processes. Electrochemical processes themselves are made up of nothing more than a myriad of tiny interactions between quantum-level particles. Thus, reduced to their most fundamental level, brain functions depend on quantum-level processes. And, since quantum processes by nature include at least some nonlocal phenomena, many of my scientist friends think it is reasonable to believe that quantum nonlocal interaction in the brain might be involved in extrasensory perception.

However, not everyone is happy with this comparison. Many of ESP's critics have argued that nonlocality only works at a very small scale—among particles so tiny they can never be "seen" by either human eyes or human-built instruments. The same folks argue that nonlocal interactions also only happen in very cold environments—say, approaching the temperature of deep space. Recent discoveries, though, have weakened these objections by showing that nonlocal entanglement can happen in large scale systems at room temperature.[4]

4. See, for example, "Living in a Quantum World" by Vlatko Vedra, in Scientific American, June 2011, pp. 3843, which discusses entanglement experiments over various temperature and size ranges.

There are still other problems with the idea of entanglement as an explanation of ESP. It seems unlikely that physical nonlocality can transmit the kinds and amount of information that we often find in remote viewing and other ESP results. If that is so, it doesn't matter how large a scale the phenomenon can extend to or how high a temperature such relations can tolerate. More importantly entanglement, as presently understood, cannot explain how experiential information (that is, how it 'feels" at a distant location, what "smells" one can experience, and so on) can be retrieved through time, whether past or future.

What all this amounts to is that some non-locality principle may someday explain remote viewing, but on the other hand, there seems an equal or better chance that it may not. So don't get your hopes up.

Quantum nonlocality works its magic regardless of distance or time - it even works on a galactic scale. Some people think the lack of time and distance constraints make ESP similar to nonlocality.

But if nonlocality doesn't ultimately explain how remote viewing works, what else is there? Maybe nothing in the physical world. I'm not talking here about spooky or spiritual things as a solution. (Though, who knows? Maybe there is a "ghost in the machine" that we'll discover some day.) Instead, I have in mind a different model. Perhaps consciousness and its effects, such as ESP, constitute a force or field that exists independently of the physical world. Consciousness could be a primeval force existing side by side with physics, rather than being a product of physics the way mainstream science thinks it is.

Many are scandalized by this possibility. They worry that it moves human consciousness and ESP outside the realm of physics and science. This would mean that consciousness might never be fundamentally explainable in scientific terms. Instead, it would be, as philosophers are fond of saying, a "brute fact" of the universe which we could never fully understand, at least in physical terms.

This fear, it seems to me, is irrational. If we discover that is the way the universe really is, we should be prepared to accept it. It would do us no good, and probably some harm, to go on pretending such a phenomenon didn't exist. What if it does turn out that consciousness is not a product of physical principles? Obviously we are still able to interact with ESP in the physical world, so there must be a connection between the two, even if only through some kind of "conversion" function, or bridge forces or principles — a way to translate some aspects of conscious experience into physical terms, and vice versa. And contrary to the fears of many scientists, even if ESP and consciousness turn out to be independent of physics, that doesn't mean science still can't learn a lot about it merely by mapping its boundaries and watching how it interacts with the physical world. Researchers already do something analogous to this in experimenting with things in physics that can't be directly observed either, such as subatomic particles and distant interstellar phenomena—or quantum nonlocal interactions.

In the meantime, I think we should continue to look for an explanation for ESP in the physical sciences. Even if we don't find a final explanation, we can certainly still learn many important things about ESP and its relevance to human nature in the process. But we should be prepared to find out in the end that there is no satisfactory explanation in the physical world, just in case there really isn't one to be found there. Which is, of course, exactly how our ancestors coped with physical phenomena before scientists found out everything that we now know about the world.

To believe that the universe is made up of nothing more than physical things requires just as much faith, and is no more provable, than the possibility that some elements of the universe, such as consciousness or ESP, are at least partly nonphysical.

Chapter 11
The Evidence for Remote Viewing

Persuading the uninitiated of the legitimacy of remote viewing has never been easy, especially when hidden behind a wall of government secrecy. It was 1973, and the Stanford Research Institute had spun off from the university it was named after and was in the process of rebranding itself as SRI International. Hal Puthoff and Russell Targ, now well into their unique government contract to look into military uses for ESP, had some exciting experimental results and were eager to go public with what they had found. But there was a catch. The paymaster on the contract was the Central Intelligence Agency, and SRI was pledged to secrecy about it. The fact that SRI was experimenting with ESP was not a secret, but the identity of the client was.

Remote viewing researchers Russell Targ and Harold E. Puthoff at the front door to SRI-International, the massive think-tank where they carried on their research.

Puthoff and Targ made their case to the CIA: The scientific community already knew what SRI was up to. Not publishing the results of their work would look highly suspicious to their fellow scientists. The logical conclusion for all that secrecy would lead straight to the keeper of all secrets, the CIA. It was a plausible worry. To show how easy such a connection was to make, my future wife used the same reasoning to figure out what I did for the Army when we were first dating. I was forbidden to tell her anything about my remote viewing, but all she had to do was scan my book shelves to divine the one subject that I never talked about. I was shocked the day she said "I know what you do for the military — you work with ESP!" I wonder to this day which others of my friends made the same deduction, but never admitted it to me.

Puthoff and Targ were correct—secrecy in the scientific community always raises suspicions about the kind of client one is working for. So in the case of remote viewing, they argued that publishing results from a secret program would actually protect the secrecy of the espionage connection. The CIA grudgingly gave its permission, so long as the government connection stayed hidden.

Getting the CIA to agree turned out to be the easy part. Persuading the scientific community to listen was harder still. When you're dealing with something as controversial in the scientific world as ESP, getting even the best results published in a professional journal can be much like getting a pack of well-fed wolves to eat a pile of carrot greens. Puthoff and Targ made their first foray with a relatively short article in the prestigious journal *Nature*. Their article detailed some of their preliminary findings working with the controversial psychic Uri Geller, who was famous for bending silverware on the nightclub circuit. Under controlled conditions the SRI researchers successfully documented Geller's ESP abilities, but could not confirm his mental metal bending. The article in *Nature* led to a firestorm of indignation from many of its readers, offended that the journal would publish anything on such a disreputable topic as ESP.

Within a couple of years Puthoff and Targ had amassed so much additional interesting data that they needed to do more with it. Both the SRI scientists were established members of the Institute of Electrical and Electronic Engineers. Puthoff had a masters in electrical engineering and, after a stint in Naval Intelligence working at the National Security Agency, earned a physics Ph.D. from Stanford. Targ had a masters degree in physics and early patents in laser design. They had previously each independently published articles on laser physics and related subjects in the IEEE's highly-regarded journals. So now working together, they approached Dr. Robert Lucky, the editor of *Proceedings of the IEEE*. Lucky's day job was Director of

Communications for Bell Labs. Neither Puthoff nor Targ had ever met him, but he was willing to accept the paper for peer review. Two of the anonymous reviewers came back with positive responses, but the third replied that "This is the kind of thing I wouldn't believe in even if it were true!"

Puthoff and Targ then offered to present their data to an audience of Bell Lab's engineers. If the engineers liked it, then IEEE would feel better about publishing the paper. The presentation seemed to go over well, and at the close of the talk the engineers engaged in lively conversations amongst themselves trying to come up with a theory about how remote viewing might work.

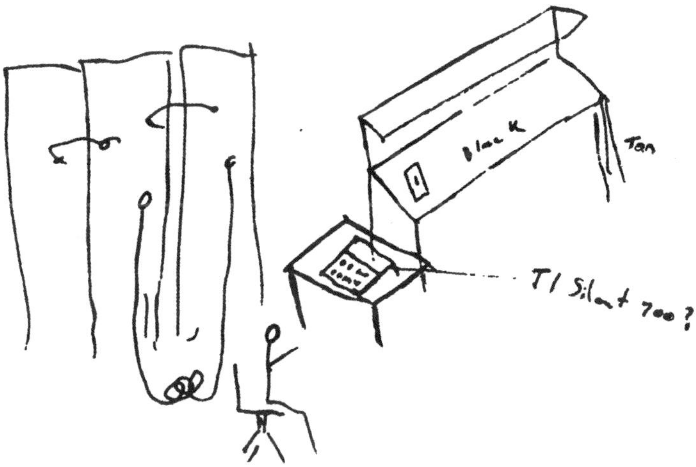

One of the experimental results published in the IEEE paper. The target was a computer terminal with relay racks behind it. The viewer even attempted to identify the brand and model of the terminal, though that was incorrect.

Dr. Lucky was still understandably skittish, so the two SRI scientists finally struck a deal with him: Lucky would conduct his own informal remote viewing experiment in-house under scientific controls, and if the experiment succeeded the IEEE journal would publish the paper. The experiment, using Lucky's interested secretary as remote viewer and a properly–blinded target set, was done without assistance from anyone at SRI. Lucky was impressed enough with the results that the Puthoff/Targ remote viewing paper appeared in the journal's March 1976 issue. In later issues the Proceedings published several commentaries on the paper, and other articles on scientific examination of ESP as well, including those from an IEEE-sponsored symposium on the subject. As you can see from this story, the best witness for the defense in the case of remote viewing is direct scientific evidence. But the notion of scientific "evidence" is often misunderstood. People frequently demand, "I want to see the scientific proof." The fact is, only mathematics and the legal profession regularly deal in proof; science rarely "proves" anything definitively, but rather only provides greater or lesser evidence for a claim. This is why the scientific world at one time may firmly believe something (for example, that Earth is the center of the universe), and at another time believe something else altogether different (that the Sun is the center of a Solar System stuck out in some small corner of the universe).

Science has not changed in this respect; witness how as recently as the 1970s science believed that adult human brains no longer developed new synaptic connections. They reached a fixed state and that was it. But new discoveries have since proved that belief dead wrong. Brains continue to form new connections throughout life—a discovery with major implications for everything from the science of learning to stroke recovery. Even more recently, theories about asteroid collisions with Earth have replaced earlier theories as reasons for the dinosaur extinction. Philosophers of science often point out that no contemporary scientific theory is ever completely true or accurate; a theory that science believes now is likely going to change or be replaced as we learn more about the world.

Fortunately, the evidence for remote viewing (as well as other ESP phenomena) is now very strong. But you will not hear me say remote viewing has been "proved" or that something counts as "proof" that remote viewing is real, any more than you will hear honest scientists say that "quantum mechanics is proved," or that there is "conclusive proof" that evolution happened in a certain way. Though the evidence may be strong that some phenomenon or scientific claim or other is true — even so strong that it verges on proof — there is always the chance that something new and more definitive may turn up. So what kinds of evidence support remote viewing?

Anecdotal Evidence

Stories and accounts related by people make up what is called "anecdotal evidence." Contrary to what some critics imply, anecdotal evidence is indeed evidence that has a certain, if limited, value. However, it is the least reliable kind of evidence. It requires trust that people are telling the truth about what they experienced or saw, that their memories are accurate, that they are good observers, and that the remote viewing experiences they relate were not contrived to deceive those observing. In any scientific field (not just the study of ESP) there is always the chance that any of these things might be wrong with anecdotal evidence. So anecdotal evidence *is* evidence for ESP, and some of it might even be a hundred percent correct. But because it can't be checked (by definition, it "already happened"— you can't go back and watch it happen again), and human nature is such that we can't trust humans to necessarily be accurate reporters, it is not very *strong* evidence for ESP.

Statistical Evidence

Generally, the most rigorous and scientifically desirable form of evidence is statistical. If the statistical procedures are correctly used, this is the kind of evidence that is the least affected by human weaknesses (such as biases, preconceptions, mistaken perception, faulty memory, poor judging ability, etc.). Statistical evidence allows for results to be measured; you can assign numbers to results. In turn, these numbers can be compared to the numbers generated by control groups, or from other experiments. I'll say more about statistical evidence for remote viewing in the next chapter. Now, though, I'll show you another form of evidence for remote viewing—evidence that falls somewhere between entertaining anecdotes and scrupulous statistics.

Prima Facie Evidence

This midrange of evidence is named from the Latin for first (prima) and face (facie), literally "at first appearance." This is evidence that is so obvious on its face that it is practically indisputable, so long as no better explanation can be found as to what caused it. In the case of remote viewing, it would be a result that was produced under rigorous conditions that yielded an outcome so clear that it would be implausible that any non-ESP cause could explain it. With prima facie evidence, you don't need to rely as much on someone's memory or trustworthiness because anyone can examine the actual result and see that the match between remote viewing data and the target is obvious.[1]

However, this is still not the most desirable evidence because it depends on our ability to correctly interpret what we see. Prima facie evidence can't be "measured" in a meaningful way when it comes to remote viewing. It can only be compared to the actual target (or photographs of the target) to see how well, in our judgment, the results match the target. Rarely can such results be analyzed statistically or mathematically, both of which are important to weed out unintentional bias or misinterpretation.

Still, prima facie results are usually so impressive that a rational person will agree that there is a clear and definite match, and that the remote viewer has correctly described a target of which he or she had no prior knowledge. Of course, to be credible, prima facie evidence has to be the result of a clean scientific process that makes it impossible for the viewer to get information from any non-ESP source. You will see some of this prima facie evidence below and elsewhere in this book. You can make up your own mind whether it validates remote viewing.

There is space here to show you only a few excerpts and sketches from remote viewing sessions of the many that have been done. Some are from military remote viewers and others from some of my remote viewing students. On the following pages, you will see sketches the viewers did while fully blind to their targets. Next to the sketches will be photos of the targets that were revealed to the viewer after they finished their sessions. And, just so you don't think that these results were cherry-picked from otherwise inaccurate remote viewing sessions, I will post the entire transcripts from the sessions, along with some commentary, on this book's website. (Online at http://www.guidetoremoteviewing.com)

---

1. Really good prima facie remote viewing outcomes are often called "8 martini" results. This is because, according to one Customs agent speaking on ABC's Nightline program in 1995 who used remote viewing data to capture a fugitive, these results are so good that it takes eight martinis to calm down after seeing them.

Example 1: Hagia Sophia

The sketch below came at the end of four pages of increasingly accurate sketches and verbal impressions. As is the case for all the examples that follow, the viewer was fully blind to this target and had no external input that could have introduced any information, much less in this much detail. At the bottom is a photo of the intended target, the Hagia Sophia church/mosque in Istanbul, Turkey. Here, as in the following pages, the target was revealed to the viewer only after all work was done on the session and the data was a matter of record.

(photo courtesy www.dailyistanbultours.com)

Example 2: Blue Nile Falls, Ethiopia

Shown here are the results from just two of several different student sessions I have in my archives with the Blue Nile Falls in Ethiopia as the target. Both sessions were performed double-blind, but six years apart (one in 2006, and one in 2012) in two different locations, by two different viewers separated by hundreds of miles. Neither viewer has ever met or communicated with the other, yet their results accurately reflect ESP-based perceptions of the intended target.

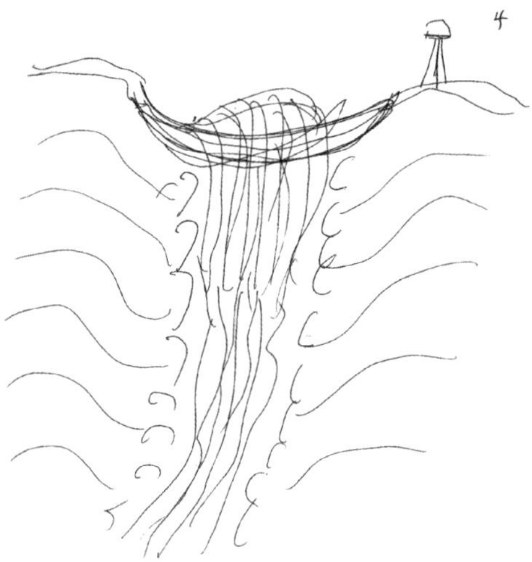

6
hard
rough
rocky
roaring
slapping
AOL BK
waterfall
over rock

Example 3: Brandenburg Gate, Berlin

(Norbert Aepli/Wikimedia)

Example 4: Western ("Wailing") Wall, Jerusalem

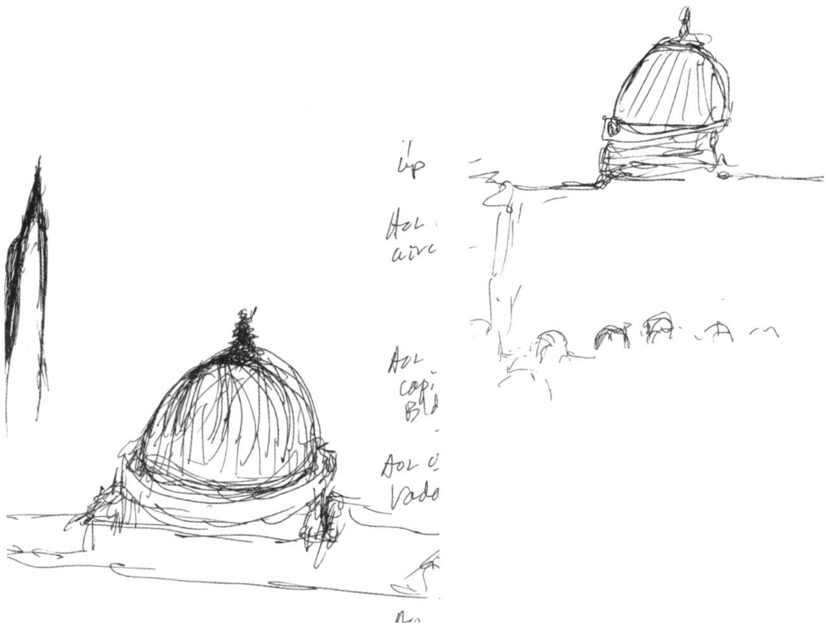

Example 6: Meroe Burial Pyramids, Sudan

For this example, I have included the results from three different viewers in independent sessions done against the same target, but months and even years apart. This shows you that even if the way they produce results may differ from viewer to viewer, the signal comes through very much the same.

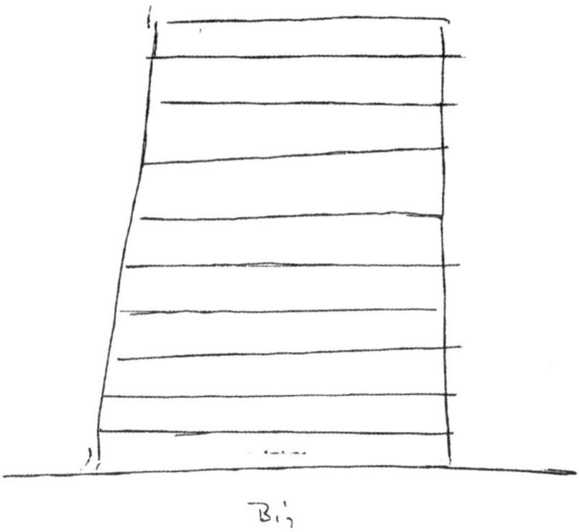

Viewer 0362 (2003)

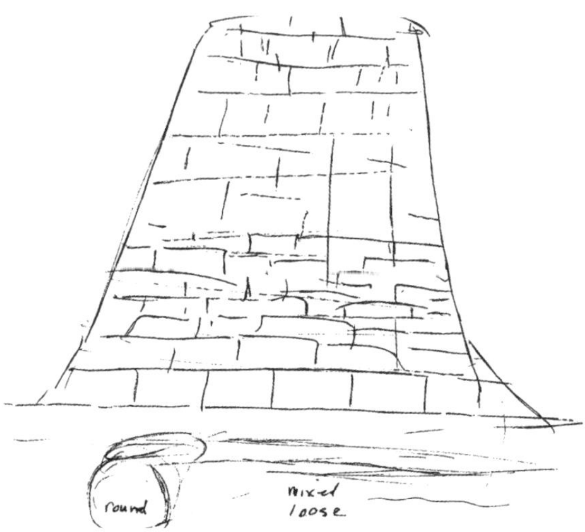

(Lassi/Wikimedia)

(Petr Adam Dohnálek/Wikimedia)

(Viewer 0465 (2004)

(Viewer 0796 (2007)

Example 7: Nuclear Power Plant

In this example, you see a different mode of presenting remote viewing results. After going through several pages of sketching and verbal descriptions, here the viewer built a three dimensional model of his impressions.

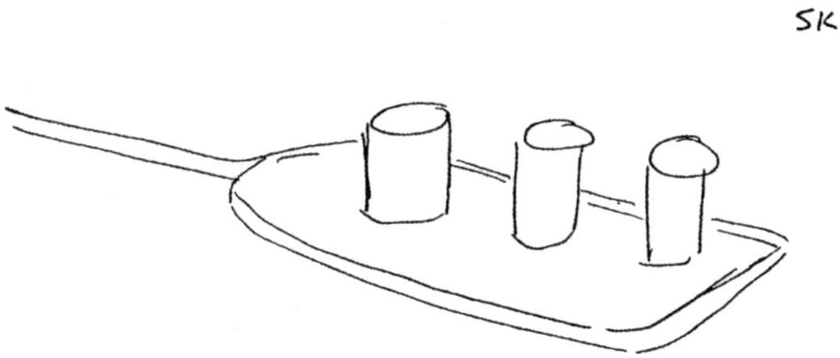

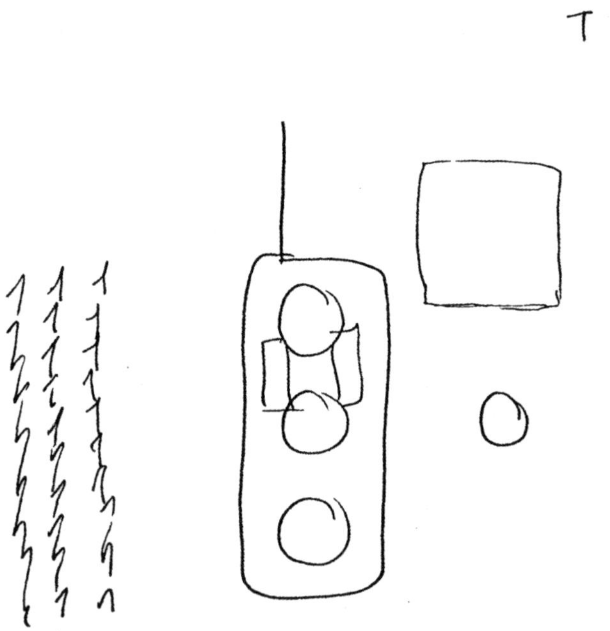

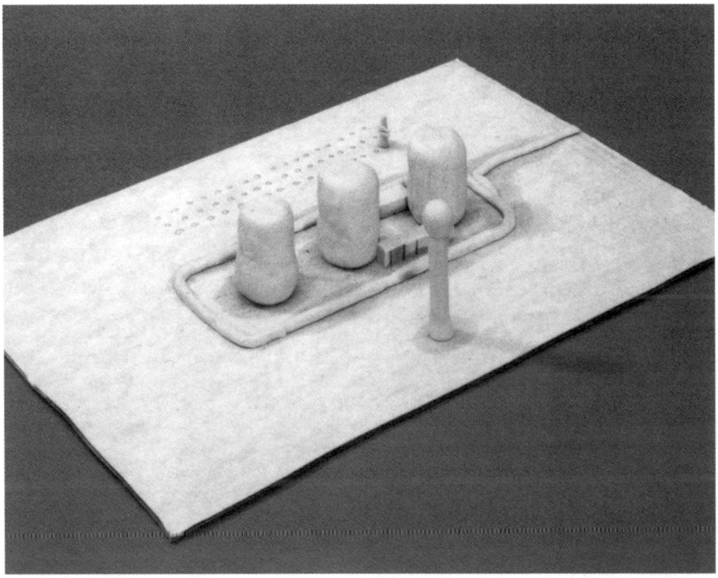

The example on these pages are just a few of thousands of remote viewing sessions that have been done over the years. You see here just a handful of sketches from among the many good ones. To be sure, not all remote viewing results match their targets as well as these. This shouldn't be surprising, as there are thousands of people working with remote viewing, and their abilities and experience range from experts with decades of involvement all the way down to rank beginners. Results run the spectrum from remarkably good matches, to those with fair similarity to the target, to only marginally good data, to almost no match at all. Many successful sessions have excellent written descriptions, but the sketches are not particularly striking. And then there are plenty of sessions which yield excellent verbal descriptions and good sketches.

Chapter 12
Remote Viewing: Evidence of ESP

Your mind can roam far beyond your body. You can perceive distant or hidden places through mental powers alone. You can develop the ability to perceive real events both forward and backward in time.

Ridiculous? Absurd? Without good evidence, such claims sound like they come from frauds, charlatans, or simple fools. But as I promised in Chapter 11, there is scientific evidence for the statements above. In this chapter I will deliver on that promise with some fairly detailed statistical evidence.

One value of statistical analysis of remote viewing is that it allows results to be measured and compared. This has several benefits. In many cases, subjective judgment (a human being looking at a result and deciding what it means) makes it appear that there has been a real effect, but a statistical analysis of the same evidence shows that the outcome was just a fluke, or a matter of imagination, or mistakes.

On the other hand, statistics can also help confirm that a real result has taken place. Confirmation is important because, among other things, it is added evidence that our subjective judgments about an experiment's results are correct. Even more exciting for

Statistics professor Jessica Utts, whose work has made a strong case for the legitimacy of remote viewing and other remote perception evidence. (Courtesy, Jessica Utts)

remote viewing researchers, statistics can also show that there is a successful result or that at first wasn't obvious through subjective analysis because of the "noise" level in the experiment.

Another possible use for statistics in remote viewing may be to compare different experiments to see whether one technique works better than another. Also, statistics can help measure improvement in viewer performance or in success against various kinds of targets. Good statistical analysis may spell the difference between guessing whether a certain approach is any good, and knowing that it is.

Point of clarification: Don't get confused about what statistics really are. People are often intimidated by statistics in general; they don't understand how statistics relate to the real world. Laymen often think of them as abstract numbers and formulas that don't clearly reflect reality. You will have an easier time if you remember that statistics are a way to keep track of a lot of real world events and plot how they connect to one another. Election statistics, for example, are nothing more than an

The Significance of Statistics in Mind-Matter Research

JESSICA UTTS

Division of Statistics, One Shields Ave.
University of California, Davis, CA 95616

Abstract — Statistical methods are designed to detect and measure relationships and effects in situations where results cannot be identically replicated because of natural variability in the measurements of interest. They are generally used as an intermediate step between anecdotal evidence and the determination of causal explanations. Many anomalous phenomena, such as remote viewing or the possible effect of prayer on healing, are amenable to rigorous study. Statistical methods play a major role in making appropriate conclusions from those studies. This paper examines the role statistics can play in summarizing and drawing conclusions from individual and collective studies. Two examples of using meta-analysis to assess evidence are presented and compared. One is a conventional example relating the use of antiplatelets to reduced vascular disease, and the other is an example from mind-matter research, illustrating results of ganzfeld and remote viewing experiments.

Keywords: statistical evidence — *p*-values — meta-analysis — repeatability

One of Dr. Utts's most influential peer-reviewed papers, which demonstrated that the evidence for ESP is much stronger, for example, than the statistics showing that a daily dose of aspirin helps prevent heart attack. (Journal of Scientific Exploration)

accumulation of thousands or millions of cases of individual people going to polling places and casting their vote, which are then evaluated according to the various choices a voter can make. The resulting statistics make it easy to see what large numbers of similar events—many instances of individual people voting—can tell us about the overall picture.

In remote viewing it is usually easy to examine one session and its results and get a general idea about whether it was a success or failure, based on the eventual feedback (such as the viewing of Shipyard 402). But to plot a thousand such sessions and have confidence that remote viewing is more than a one-hit-wonder, we need a tool to keep track accurately. Statistics provide that tool.

The simplest way to illustrate this is to examine one small-scale set of remote viewing results that were analyzed statistically and explain what the statistics mean. For that, I'll use a series of three unpublished pilot experiments in which I worked as the viewer. These were associative remote viewing (ARV) experiments, which is a mode of remote viewing I explained in Chapter 6. It is enough just to remind you here that in ARV the viewer is trying to describe an object (for example a photo) that he or she will be shown at an appointed time in the future. That object is then serves as a signal as to how a specific future event or outcome will turn out.

For these three studies, the results were as follows:

Experiment A (March through June 2002): 19 remote viewing attempts, yielding 14 correct, 3 incorrect, and 2 passes.[1] Statistical analysis yielded a p-value of $p = 0.006$, or only 1 chance in 167 that the results could be that good on accident. (This is a statistically significant outcome).

Experiment B (August through December 2002): 39 remote viewing attempts, yielding 22 correct, 15 incorrect and 2 passes. $P = 0.16$, or 1 chance in 6 that the results were accidental (positive, but not statistically significant).

Experiment C (October through December 2002)[2]: 14 attempts, 8 correct, 3 incorrect, and 3 passes. Results yielded $p = 0.11$, or 1

1. A pass occurs when the results are so ambiguous that no decision can be made with any level of confidence. For statistical purposes, passes are often not counted, and totals may be adjusted accordingly.

2. Note that experiments B & C were run partly over the same several week period. The same viewer was involved, but the researchers were different (and unaffiliated).

chance in 9 that the results were accidental (positive but not quite statistically significant).

All three experiments taken together yield a statistically significant result of p = .003, or 1 chance in 333 that the results were produced by accident. Experiment B produced the weakest results, though they were still noticeably positive in the same direction as were the other two experiments.[3]

I'll assume that readers of this book are novices when it comes to statistics. I was (and still am), too. But my Ph.D. course work forced me to get a better grip on basic statistical analysis. So I will act as translator here. Let's take Experiment A as our example. In this experiment, I made 19 attempts to correctly remote view which one of two photos I would be shown a day or two after I made each attempt.[4]

Of those 19 attempts, I described one of the two photos in each attempt well enough that the judge picked the correct target 14 times. Three times we got it wrong, and two times the judge took a pass—couldn't decide which photo I was describing. So that means (counting the passes as if they were misses) I got it right 14 out of 19 times. If you were simply guessing between two photos 19 times, by chance you would get it right only half the time, or 9 1⁄2 tries out of the 19.[5]

So being right 14 times is pretty good compared to the 9 1⁄2 that we should expect from a random outcome. (And it is even bet-

---

3. Notice that experiment C was also not statistically significant, yet the odds from all three experiments together against the result being an accident were even higher than the scores of any of the individual experiments. How can it be that the results of all three together turn out better, even if two of them weren't significant? The reason is that the results of the two non-significant studies were still greater than chance, so they do add to the total number of successful trials.When you add them all up, they increase the significance of the overall score. Since the trials in all three of the studies each involved 50/50 chances, you can lump them all together, and calculate the probabilities from there.

4. Each remote viewing attempt had a two-photo set that was different than all other photo sets in the experiment, so there were 19 different photo pairs for the 19 remote viewing sessions.

5. Obviously you can't get a choice half right, so if we were actually guessing we would end up with 9 or10 out of 19.

ter if you take the passes out of the calculation and do the statistics on 14 correct out of 17 times.)

According to the laws of probability we can occasionally get a result that good purely by luck. But how lucky would we have to be? Statisticians measure that by assigning a "p-value." In the case of my Experiment A, "p = 0.006" puts a number on how likely it would have been, just by accident, to get 14 right out of 19 instead of the regular 50/50 chance which would have yielded only 9 1/2 correct answers. The p-value is the shorthand statisticians assign to show how far the result diverges from simple chance.

For nonstatisticians this number can be a little hard to grasp. Thankfully, we can use a simple statistical formula to convert the p-value to something more easily understood. A p-value of 0.006 is the same thing as saying that one would get 14 correct out of 19 on average only once in 167 tries (or as it is commonly said: "one chance in 167" of getting this result).[6] If we were flipping 19 pennies trying to get all heads, we would on average have to do 167 sets of 19 flips (3,173 flips total!) to get at best 14 heads in only one of those 167 sets. (The number 167 comes simply from "doing the math" of the number of possible outcomes when factored with the number of attempts.) The odds of getting all 19 to be heads is even greater.

Scientists generally say that a result reaches "statistical significance" when it has a p-value of 0.05. Put another way, the results are considered significant if there is only a 5 percent chance, or 1 chance in 20, that the result from that particular experiment happened by accident. The bottom line is that the more zeros you have to the right of the decimal point before you get to a number, the less likely that a given outcome will have simply been happenstance. More than one zero is good evidence that something beyond luck may be at work. Our p-value in Experiment A is .006. So if something is statistically significant at .05, then result for Experiment A with its two zeros might be considered highly statistically significant.

---

6. The statisticians reading this will realize this is not exactly right. As I mentioned before, this is because the passes were not counted in these statistics, since they are essentially neutral. The actual success rate is 14 out of 17 trials for statistical purposes.

Statistical analysis is often much more involved than this, with complex formulas and principles available for experts to use in a variety of situations. When more than two outcomes are possible, such as in a horse race, for example, the statistics get more complicated as well. But the basic principles themselves remain the same. With this background, you should be able to understand the gist of the following statistical results.

I will first list a summary published in 1984 of remote viewing experiments conducted up to that time. Researchers found 28 published studies. Fifteen, or more than half of these studies, were statistically significant. This is an impressive total, since if mere chance were at work, we should expect just under 2 (actually, 1 1/2) out of those 28 experiments to have turned out statistically significant by accident.[7] Obviously 15 successful experiments out of 28 total far surpasses that.[8]

Another series of remote-viewing style experiments was done by the Princeton Engineering Anomalies Research (PEAR) lab at Princeton University. These were called "precognitive remote perception" (PRP) experiments, and differed from the SRI remote viewing experiments in one important way. Like the SRI experiments, the viewers were blind to the targets and separated from them by distance. Unlike the SRI work, the PEAR experiments viewed targets in the future. PEAR conducted nearly 1,000 PRP trials in the 28 years it was in business. Of these experiments, 653 were formal trials that were statistically analyzed.[9] The resulting anal-

---

7. This is because probabilities work not just in the results of individual experiments, but they apply on a larger scale in groups of experiments as well. Just as it is possible that every so often the results in an experiment can turn out statistically significant just by accident alone—after all, on rare occasions someone flipping coins might accidentally get ten heads in a row—it is unlikely but still statistically possible that one out of 20 experiments might turn out significant at the p=0.05 level just by chance. (However, if there really were no ESP effect at work, it would still be a surprise if any study turned out significant.)

8. The same survey found 18 unpublished studies in addition to the ones that were published. Eight of these unpublished studies were also statistically significant. This has implications for the so called "file drawer" problem (when scientists publish only successful studies and file all the negative ones away, deceptively making the statistics look better than they are). The remote viewing "file drawer" includes both positive and negative study outcomes, suggesting that there wasn't any kind of file drawer problem.

ysis showed that the PRP experiments were highly significant at p= 0.0000000003 (notice all those zeros). There is only one chance in 300 million that such a result would happen on accident.

A secret military study completed in 1989 calculated even more impressive statistical results. Congress ordered the Defense Intelligence Agency (DIA) to do a thorough review of the military's remote viewing research from its start in 1972 through the end of 1988. DIA turned to the scientists at SRI for a thorough evaluation. Their review showed that 25,449 trials were done under scientifically controlled conditions in a total of 157 experiments by 227 remote viewers. The results were calculated to be $p = 6.12 \times 10^{-14}$. This is one chance in 16.3 trillion that the results came by accident, making it staggeringly unlikely that ESP was *not* involved. There is one more experiment that I will tell you about. It involved ARV.

Christopher C. Smith, lead researcher in the successful University of Colorado student experiment in associative remote viewing.

---

9. The remaining 300 or so 'informal' trials were exploratory in nature, or not as tightly controlled, and therefore did not reach the PEAR standards for inclusion in the overall statistical evaluation, though their cumulative success ratios would have made PEAR's results even more impressive.

If You're Psychic, Then Why Aren't You Rich?

I used this question earlier in the book to introduce the idea of making money using remote viewing, and briefly mentioned the experiment I'm now going to describe in detail. Here are more of the facts: In 2008 my son Christopher took a course at the University of Colorado from physicist Garret Moddel. Dr. Moddel taught the class as an introduction to the science of parapsychology, and invited me to one of his class periods to give an overview of remote viewing. I introduced the students to associative remote viewing (ARV), which you now know is a simple way of using remote viewing to predict future events.

As an exercise we predicted the winner of a basketball game between the Atlanta Hawks and the Chicago Bulls that wouldn't end until well after the class made its prediction (though we didn't reveal which teams were involved until after the viewing was done). According to the ARV results, twelve of the attending 17 students and faculty predicted the Hawks would win. When the game was over, the Hawks had beat the Bulls 113 to 108. With odds expected by chance of only 50/50 this is a fairly impressive outcome, with a pvalue of about p=0.07 (or slightly more than one chance in 14), which is just shy of statistical significance.

Flush with success, the students decided as a class project to use ARV to predict the closing numbers of the Dow Industrial Average. A benefactor kicked in $10,000 from his investment funds so real money could ride on the outcome. The students contributed 67 sessions, of which 54 were hits. This result was statistically significant at p <0.01, or odds against chance of less than 199-to-one that this would happen by accident. But, as Dr. Hal Puthoff likes to say, "What really counts is money in the bank." Using the ARV results, the investor made seven straight correct trades. When the smoke cleared his $10,000 had grown to $26,000, for a net gain of 160% on the investment.

I'll mention Greg Kolodziejzyk's ARV project again here, in terms of the statistics (and not just the $150,000 profit) it generated. Greg, a computer scientist who developed a profitable computer graphics company that he later sold to Adobe, kept impeccable track of the statistics he generated over the years while he ran his ARV project. Greg ran 5,677 associative remote viewing trials during the period,

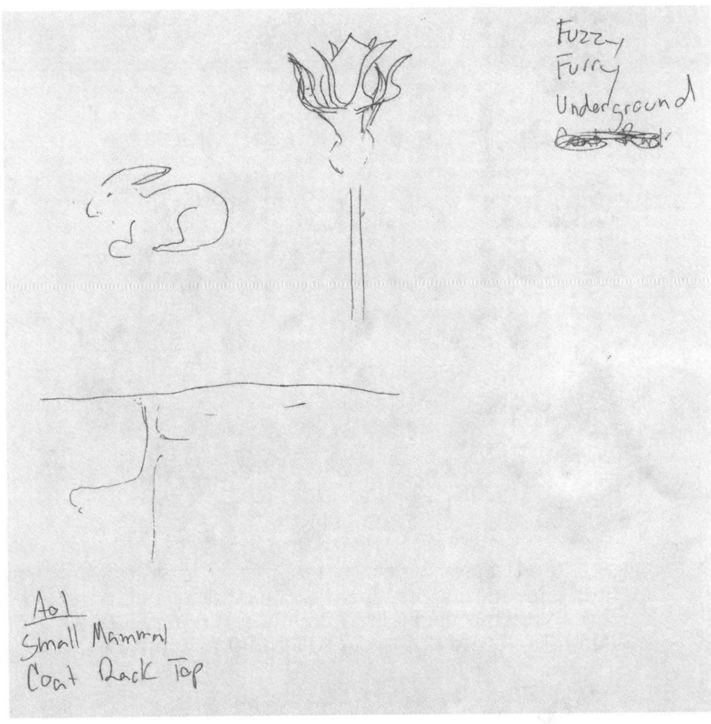

Above are the two photos in an ARV target set from the University of Colorado remote viewing project, compared to the one-page remote viewing response. The hamster photo represented the correct target choice, and the results point to that photo (comments such as, "fuzzy," "furry," "underground," "small mammal," coupled with sketches of a rabbit-like rodent and diagram of underground burrows; additionally, the sketch that at first glance looks like a thistle blossom, also resembles the claws on one of the hamster's paws).

approximately 53% of which turned out to be correct.[10] This yielded a highly statistically-significant z-score of 4.0, (another way of measuring statistical significance) or, to phrase it in the terms we are using here, p=0.0001.

There are more standard and associative remote viewing experiments whose successful results could be listed here along with their statistical significance, but I am in danger of losing your attention altogether. In the interest of moving the remote viewing story along, that is enough for now.

Greg Kolodziejzyk, who conducted the longest-lasting and one of the most successful associative remote viewing projects, involving over 5,000 individual trials and netting $150,000 in profits.

10. 53% may not seem impressive, but Greg was extremely conservative in the way he designed his experiment to reduce displacement to the wrong target from entering in. For that reason he chose not to use a number of strategies that have been shown to increase the percentage of correct predictions. He relied instead on as much computer interaction as possible and large numbers of trials to offset the inevitably weaker effect. The reason Greg could have only a raw success rate of 53% and yet still have such a highly significant outcome is because he did so many trials that even a weak effect could be strongly shown to exist.

Chapter 13
Critical Thinking for Skeptics and Believers Alike

A Cold War has been declared on remote viewing and other kinds of ESP. Parapsychology research has always been criticized by dedicated skeptics whose job it is to be, well, skeptical. But disapproval of remote viewing comes also from mainstream scientists and even from some religious people. (Someone once told me that we must be doing something right to get both atheists and religionists mad at us at the same time.) No one should be surprised by these attacks. Any time you start producing evidence showing that the way a large number of people look at the world might be wrong, you are bound to run into push back. People don't like to have their comfortable web of beliefs shaken, so when you start showing that cherished scientific truisms might be either incomplete or maybe even altogether wrong, things are bound to get ugly. And indeed they have.

The causes of this war, I believe, are easily traced. There is not enough critical thinking on either side of the conflict. You may be surprised to hear this from a remote viewer: There is not enough critical thinking when it comes to remote viewing, even by many of its supporters. It is not just the skeptics, but also many of the believers who need to get their critical-thinking ducks in a row. I'll start by taking on the believers. (Any skeptics who are reading this, your turn is coming!)

"True Believers"

It is often those who most passionately believe in remote viewing (and other forms of ESP) that hand skeptics the most ammunition against it. There are several ways remote viewers and remote viewing believers go astray to add to that ammunition. Understanding where the thinking gets sloppy will actually help to improve remote viewing success and strengthen the remote viewing field overall. But we have to get over human nature first.

By nature, human attention is drawn to what is most exciting or novel. Moreover, we want easy solutions to problems. We want quick, simple answers for complicated situations. We imagine we see patterns, meaning, or connections in random systems or situations, which later turn out to be completely wrong. And we have a morbid fascination with misfortune and calamity, so long as we ourselves aren't directly involved.

I'll treat this last point first. Think about doomsday fads throughout the centuries; or collective hysterias such as the Salem witchcraft trials, or mobs reacting in knee jerk fashion to lynch an innocent bystander they merely suspect of being guilty, or the frantic predictions of end times whenever humans approach the turn of each millennium. These things happen because of the way human psychology is put together. Our brain circuitry is tuned to pay more attention to misfortune and danger than it is to focus on more positive things. This makes a certain amount of sense. After all, it was more important for the survival of our distant ancestors that they know about the man--eating leopard or deadly epidemic over in the next village than it was to know that the apples on the neighboring tree were ripe and ready for eating (important as that might have been). Over many years of human development our mental equipment has created more circuitry to process bad news than it has good.

Any number of doomsday scenarios are hyped by sensationalizers in the outer fringes of the remote viewing community. (NASA image)

As a result, in everyday life our attention and fascination is more easily captured by stories of misfortune and disaster than by good news. This is one reason why negative, rather than positive news makes up the majority of what we hear or read in the media.

Bad news sells advertising better than good news.

It is no accident, then, that people who have only a passing acquaintance with remote viewing will often focus on the doomsday scenarios spouted by many of the most vocal fringe elements of the remote viewing world on various broadcast media or the Internet. Solar flare "kill shots," devastating earthquakes, detonations of nuclear weapons, and so on are predicted over and over again by self-proclaimed remote viewers. Even though the promised disasters never occur, there is always a willing audience to lap up the far fetched tales. And the tellers of the tales land on their feet by stretching the truth of real events to match their predictions.

This is what happens when we humans don't understand our own psychology—when we either haven't learned or don't exercise critical thinking skills. These skills were developed over centuries of painful trial-and-error as an antidote to normal human instincts that were, nonetheless, often dysfunctional when it came to evaluating ambiguous information or observations. Remote viewing has suffered from inadequate critical thinking since its public debut in late 1995. Doomsday predictions, unsupported claims about space aliens, and numerous other fantastical stories have all supposedly been revealed through remote viewing by less-than-trustworthy remote viewing practitioners. Thousands have believed these stories, which have consistently turned out to be wrong. This near-hysteria has seriously muddied the waters for a discipline that began its life as an attempt at straightforward scientific research. Though things are calmer today, vestiges of the hysteria still persist, almost two decades later.

I'm not going to discuss further any of the circus like atmosphere that grew up around remote viewing early in its public life. Those who have been in the field for awhile are aware of it, and those just discovering remote viewing for the first time don't need to waste their time (though they will no doubt encounter this noise soon enough). But think about how much harm a fantastical approach to remote viewing brings. Since none of the predicted sensational events turn out to be true, over time those who originally supported the field learn to be cynical about all remote viewing claims, and originally neutral or skeptical folks find plenty of justification in such foolishness to reject remote viewing and other forms of ESP altogether. Fantastic claims stunt the field.

But the "believing too much" mindset contributes to other problems. For starters, too many in the remote viewing community are willing to give bad remote viewing work a "pass." People are, for example, often reluctant even to ask under what conditions results were produced; whether acceptable remote viewing protocols were followed in producing some result that seems too good to be true; or what are the credentials of someone claiming to be an expert remote viewer or remote viewing teacher. Why is this? I think we often are afraid others will see us as being negative or overly critical for asking even reasonable questions.

As a result, dubious remote viewing work done under poorly controlled conditions often gets the same respect as good work done under impeccable conditions. This is unfair to the efforts put in by people trying hard to do remote viewing correctly. It also gives remote viewing a bad name overall, which makes it harder for serious outsiders to get past the noise to discover the legitimate aspects of remote viewing. Perhaps even worse, poor critical thinking also holds back progress in remote viewing performance by making it harder to know what the standards of good remote viewing should be. Essentially, bad thinking creates its own kind of "mental noise."

Here are some rules of critical thinking we all should apply:

1) If something sounds too good (or incredible) to be true, it probably is not true. We expect claims about ESP to sound at least somewhat incredible. After all, we are dealing with effects that seem to lie outside the power of current science to explain. But that is why critical thinking and asking for verification is essential in remote viewing and ESP. We need to be sure that there is substance and accuracy and not fantasy behind any surprising claim that is made. As my experience as a remote viewer has grown, I've learned that humans are capable of feats that mainstream science has thus far never imagined were possible. Ironically, I have also grown more skeptical of stories that just don't sound right, that sound too good to be true, that don't fit in with the other facts of the universe, as explained by what we've learned about ESP. I have looked into any number of these, and found too many of them to be based merely on the fantasy of the teller. The next three rules will guide you in sorting out the good from the bad.

2) Always ask to see the evidence before accepting a claim, and always ask to know how the evidence was developed and interpreted. Too often, people making a claim turn out to be trapped by their own imagination or wishful thinking. Sometimes they are even intentionally being deceptive. Since those listening to such a claim usually have no idea whether it is true, they often think they should be able trust the “expert” telling the tale. But all humans are fallible, and even honest people can be mistaken about things they think they saw or did and pass that misinterpretation on to other people. When the deception is intentional, things only get worse.

3) Always insist on seeing credentials. “Oh, I have to keep that secret!” is a redletter warning that the claimant is probably making things up. If he or she claims to have worked as a remote viewer for the CIA, or the NSA, or any other government service, demand to see official evidence. Real former government remote viewers can easily document their involvement. Those who can’t show such documents (or provide testimony from authentic witnesses) are almost certainly making their story up. If they say they learned remote viewing on their own, demand to know how. If they say they were taught by someone else, check out that person’s credentials.

This latter rule may be difficult to follow. After all, in our culture we tend to accept what someone says as true unless we have reason to believe otherwise. We may worry that by demanding evidence or substantiation from someone we will look like we are picking on them. We worry that we might appear to accuse them of being untrustworthy or lacking integrity. But in science, the principle is just the opposite, leading to rule 4:

4) Don’t take anyone at their word. It is always best to assume someone making an unusual claim might be at least mistaken until other evidence or other trustworthy people confirm that it is true. But people who are honestly mistaken aren’t the worst problem. There are dishonest people who are good enough actors to fool many. The fringes of the remote viewing and ESP worlds have had too many of these. Despite what I said above, even if someone has the right credentials, that doesn’t mean that person is infallible, or even necessarily trustworthy. They may be more likely to be correct, but even experts can be wrong, or intentionally misleading, or biased. It is always wise to get verifying evidence whenever possible for any important claim someone makes, even if they are authority figures.

Insisting on good, confirming evidence and, where appropriate, on knowing a person's credentials are the main ways for finding out what is true, and what is not. It is too easy for people to fool others in a field like this. Asking for evidence for surprising claims is not just fair, it should be the norm. A claimant with nothing to hide would not be upset by such a request. Those who do get upset when you ask for evidence usually are hiding something. Failing to check credentials and accepting people, even authorities, at their word damages the field and helps deceivers and wishful thinkers get away with whatever they are trying to put over on the unwary.

For Skeptics and Other Doubters.

I have taken on the "true believers." Now it's the doubters' turn. One of several critical thinking mistakes that disbelievers in ESP phenomena make is to assume that our society's understanding of scientific principles is complete and unchanging. I couldn't count how many times I have heard or read comments along the lines of, "Remote viewing [or fill in the blank with another ESP phenomenon] is impossible, because it would violate the laws of science." This is the same argument that has been made again and again against other important findings over centuries of scientific progress.

I'll brush past the oft-cited case of the thousand years it took to accept the theory that the Sun, and not the Earth, was the center of our solar system, and the trouble Galileo got into for trying to change the minds of the scientists of his time. Instead, I'll skip to more recent examples. As I briefly mentioned in Chapter 2, Ignaz Semmelweis discovered in the 1840s that simply by washing hands with a bleach solution before delivering a baby, a doctor could dramatically reduce the mother's chance of dying from infection. But Semmelweis's fellow scientists rejected his theory because it contradicted what they thought they knew at the time about the laws of nature. It took decades and many more unnecessary deaths before Semmelweis's discovery was finally accepted.

Earlier, in 1794, Ernst Florens Chladni proposed the idea that meteorites originated in space and fell to earth from the sky. It was many years before the science of the day was willing to accept the

In 1794 Ernst Florens Chladni came to the conclusion that meteorites came from space. Thomas Jefferson rejected the idea, saying it was more likely that scientists would lie than that stones would come from space. (Chladni portrait by Henry Adlard; Jefferson portrait by Rembrandt Peale)

idea. Thomas Jefferson, referring to claims from academics of a meteorite in America, reportedly said, "I would rather believe two Yankee professors would lie, than that stones have fallen from the heavens."

More recently, it took decades for the now-accepted idea of continental drift to gain acceptance after its introduction in 1912 by Alfred Wegener—again, because it was out of step with scientific thinking of the day.

Alfred Wegener proposed the theory of continental drift in the early 1900s; but the idea was rejected by mainstream science for more than fifty years.

Philosophers of science take for granted that scientists will almost always start off rejecting claims that violate their beliefs about nature. Scientists will do this even if there is extensive evidence that their views are wrong. It can takes years, and in some cases centuries, before science changes its collective mind. This is exactly what is happening with ESP and remote viewing.

There is much more that could be said on this topic of the recalcitrance of science. But I don't want this book to grow to be too fat. I'll dedicate the rest of this chapter to answer some of the more common objections I've run into from those unwilling to accept remote viewing and ESP as being real. If I don't address the specific question or objection you have, please feel free to email me at rvguide@rviewer.com.

I thought all this ESP stuff had been debunked long ago?

No, science has never disproved ESP. Many people have heard or believe that it has. Unfortunately, this belief is as common as it is wrong. Just the opposite is true; the evidence for ESP is stronger now than it ever has been. Years of criticism from skeptical scientists have helped parapsychologists hone their experiments to where even some skeptics have begun to admit that parapsychology research is often some of the cleanest science being done. The answer to the next question gives a sampling of the scientific results.

There is no evidence for remote viewing or other kinds of ESP.

If you have read this book from the beginning, then you already know that there is plenty of evidence for remote viewing. And, it turns out, there is plenty of evidence for other kinds of ESP as well. Let me summarize the evidence I found when researching my doctoral dissertation.[1] I will cover only three of the many ESP research models, including for contrast some of the remote viewing data cited earlier.

Presentiment: In these experiments, participants in present time subconsciously react to negative experiences that they will see in the future. As of 2009, 23 experiments were conducted. Twelve of these were statistically significant, or even highly statistically significant. Three other experiments did not quite reach significance but their results were in the positive direction. (In a run of 23 experiments, only about one experiment might be significant merely by chance alone.) These presentiment experiments were conducted by at least eight different unaffiliated labs in two countries. Since my own doctoral investigation wrapped up, several more experiments have been published, nearly all of which produced significant effects.

Direct Mental Interaction with Living Systems (DMILS): These experiments involve participants who try mentally and/or physically to influence others who are isolated from them by distance or shielding or both. At least 68 studies were conducted as of 2008. Of these, 40 were statistically significant. Another six experiments

---

1. Available through https://rviewer.com/dissertation-abstract/

were significant but had mixed results, and eight were positive but nonsignificant. This amounts to a total of 54 out of the 68 that were either statistically significant or at least in the positive direction. By chance one would expect to see only about three produce significant results. In fact, more than 13 times that number turned out significant. Among the 11 failed experiments were several that altered protocol to see how that would change the results. One would expect these to fail at a higher rate. If these protocol changing experiments were taken out of the total, the overall results would be even stronger.

Remote Viewing: The total number of replications (numbering in the thousands) is difficult to tabulate. But I mentioned one benchmark before in Chapter 12. A survey in 1984 showed 28 published studies were documented, with more than half (15) statistically significant (where at most only 1 in 20 should be significant by chance). 18 unpublished studies were found, of which 8 were statistically significant. I also previously mentioned a comprehensive review ordered by the Defense Intelligence Agency completed in 1989. This study reported 25,449 remote viewing trials in 157 experiments performed by 227 different individual participants yielding successful results of p = 6.12 X 10 is 6.12 followed by 14 zeros, or, one chance in 612,000,000,000,000 that the successes were due to accident. (The results were broken down more finely than this, but would take too much space to lay out here). Since then, additional research, most of it statistically significant, has been published, increasing the reliability of the data.

Didn't the CIA find that remote viewing was of no use?

No, it didn't. This question has to do with a study commissioned by the CIA in 1995. The 1995 study was conducted in a way so as to justify closing down the Star Gate program immediately after its transfer to the CIA from the Defense Intelligence Agency. To perform the study the CIA hired the American Institutes for Research (AIR), a think tank specializing in doing evaluation studies for the government. It was probably no accident that some of the lead researchers in the AIR's study had also been involved in a similar study in 1988. That 1988 report turned controversial when it was discovered that the investigators had tried to suppress research findings that were favorable to parapsychology and ESP.[3]

The 1995 AIR study claimed that it found no evidence that remote viewing was ever of any use to the intelligence community. However, later analysis showed that assessment to be skewed, and fundamentally unsound. In the body of their final report the review panel members themselves admit that they only examined a tiny fraction of the available evidence from the Star Gate remote viewing program. Of the more than 2,500 operational remote viewing sessions done during Star Gate's lifetime, the review committee only examined 40 sessions. If you do the math, you will see that this works out to less than one out of every 50 of the total. And of the roughly 240 remote viewing experiments done (which, as noted above, contained more than 25,000 trials), the AIR only examined the results of 10, which amounts to barely 4 percent of the research total. In other words, for each experiment they examined, there were 24 they ignored. Further, they only interviewed the five Star Gate personnel who still remained in 1995, but none of the 35 personnel who had served over the previous 16 years. Nor did the panel interview any of the program's hundreds of previous clients to see what they thought of the remote viewing results they had received.

In other words, the AIR didn't even bother to look at the vast majority of the evidence before declaring remote viewing to be "of no use." But there was one key indicator that the CIA had planned all along to "prove" remote viewing a "failure." The Star Gate program was permanently shut down on June 30, 1995. Its remaining personnel were shipped off to other jobs in the Washington, DC, area and, much like the crated-up Ark at the end of the movie "Raiders of the Lost Ark," its records were boxed up and locked away. But the AIR review that was supposed to decide whether to terminate the remote viewing program wasn't even begun until the month after the program was closed. So the CIA closed it down, then "decided" it should be closed down. Did the CIA prove that remote viewing was a failure? You decide.

But isn't remote viewing (and ESP in general) impossible because it would violate the laws of physics?

---

3. John A. Palmer, Charles Honorton, and Jessica Utts, *Reply to the National Research Council Study on Parapsychology*, (Parapsychological Association: Research Triangle Park, NC, 1988).

James Randi is thought of by many as the public face of the skeptical community. (Wikipedia)

In a word, no. It is true that remote viewing and ESP are not possible according to the rules of classical physics (also known as "Newtonian" physics), and could be thought of as "violating" those laws. But this really doesn't mean much. Many very real, scientifically-accepted effects and phenomena are impossible according to classical physics, yet they still happen. They happen because of general relativity and quantum mechanics. These two theories dominate physics, and classical physics is only a subset of them. It is not yet clear whether ESP conflicts at all with the rules of quantum physics. In fact, as I discussed in an earlier chapter, a quantum-level phenomenon known as "nonlocality" or "quantum entanglement" may be a possible explanation for ESP effects.

Even if it turns out that ESP is not explained by quantum physics, but instead by some non-physical process, then ESP still wouldn't violate physics principles. It simply wouldn't depend on physical laws to make it real. Neither ESP nor physics would need each other to be able to exist. In that case, saying that one makes the other impossible or "violates" its laws simply doesn't make any sense.

Positive ESP results are due to fraud.

Is fraud really widespread in ESP research, as is claimed over and over by skeptics? The answer is no. Anytime a skeptic is challenged to provide evidence for this claim, he or she will always point to the same handful of cases (out of the thousands of ESP experiments completed over the years). Interestingly, skeptics have rarely actually proved fraud in a parapsychology experiment; many frauds have been uncovered and reported by other parapsychologists. Given the scrutiny on parapsychology experiments, and all the known ways experiments can be fudged (as happens at least as often in mainstream science), it would be hard to perpetrate a fraud anyway.

Hidden behind all those charges of fraud, I suspect, is really the mistaken belief that all the practitioners of so called "esoteric arts," such as palm readers, fortune tellers, astrologers, and so on, are part of parapsychology. There is undoubtedly fraud and deception that goes on in the esoteric arts community, even if many or most of the practitioners are sincere. But none of those folks, fraudulent or well-meaning, are parapsychologists. They do not count as scientists, and they seldom if ever participate in scientific research. To count them along with real parapsychologists is a category error—a logical fallacy—that many skeptics make.

So here's the bottom line: While there may be fraud behind many "psychic" claims, fraud in actual parapsychological research is really quite rare. To be sure, some positive ESP findings are the result of poor research design or methods. In any field of science research, poorly designed experiments lead to false positives. The opposite can also be true - poor research design can lead to false negatives as well. There have been poorly constructed ESP experiments which showed results that weren't really there, and no doubt some that were so poorly done they stifled real effects. But one of the blessings of close skeptical scrutiny is that flaws in ESP experiments are pointed out and corrected, often more quickly even than in mainstream science. Several important research methods in use in science today were either developed or improved upon to aid ESP research or make it more accurate and effective.[4] The quality

4. Some examples: the electroencephalogram (EEG), various improvements on statistical analysis (such as meta-analysis tools), randomization procedures and tools, and double-blind testing procedures.

of parapsychology experiments has improved to the point that in the AIR review of remote viewing, Dr. Ray Hyman, the lead skeptical researcher could find no flaws in the experiments he was critiquing, and in fact admitted that he could find nothing that would account for their positive effects. Nevertheless, he refused to admit that ESP might actually be happening.

Leading skeptical scientist Ray Hyman admitted that he could find no flaws in scientific experiments that showed strong evidence of ESP results. (Wikipedia)

Any positive ESP effects that aren't the result of fraud or flaws are the result of confirmation bias.

Confirmation bias is a typical human psychological trait that simply means the tendency to see what you want to see in a situation. A version of this is called selective validation, which means giving more weight to evidence that supports your belief and less weight to evidence that doesn't. In the case of remote viewing, a simplified example of selective validation might go something like this: During a remote viewing session, a viewer produces 15 data points. Three of them vaguely match the target, while 12 of them don't match the target at all. The viewer declares the session to be a "hit," pointing to the three items that somewhat match, while downplaying or ignoring the 12 that don't. Selective validation can take other forms, but in principle differs little from this example.

It is possible for selective validation to occur at any level in remote viewing and ESP research and practice. But experienced remote viewers and the scientific parapsychology community have for decades known about the confirmation bias and selective validation traps. Though most skeptics seem uninformed about this, considerable effort has been made in parapsychology research to adopt standards and protocols that avoid these problems. Seldom is it possible these days for either selective validation or more general confirmation bias to play any role in producing the positive results of legitimate ESP and RV experiments.

Ironically, skeptics themselves often fall prey to confirmation bias. As one example, in 2003, I gave one of my many talks involving remote viewing. The man assigned to give a response to my talk happened to be a skeptic. He had found an example on my website of a remote viewing result, along with a feedback photo of the site the viewer had been blindly targeted against. This was, he claimed, a prime example of confirmation bias, as the target and the results only vaguely matched. I have included this example on the next page so you can decide for yourself whether you think he was right about confirmation bias being involved. In my view, this is a legitimate "hit." It seemed to me then that by denying the obvious similarities, my skeptic friend was showing a confirmation bias of his own against the possibility of ESP being real.

I have by no means discussed all the possible skeptical arguments, just a few that I have heard the most often. There are plenty more where those come from, but it would take several more chapters to respond to all of them. At least now, if you were dubious about remote viewing, I hope you have more perspective and realize there are two sides to the story. If, on the other hand, you believe ESP to be real, but have had skeptical friends or relatives use these arguments on you, you know how you can respond the next time that happens. Or simply invite them to read this book.

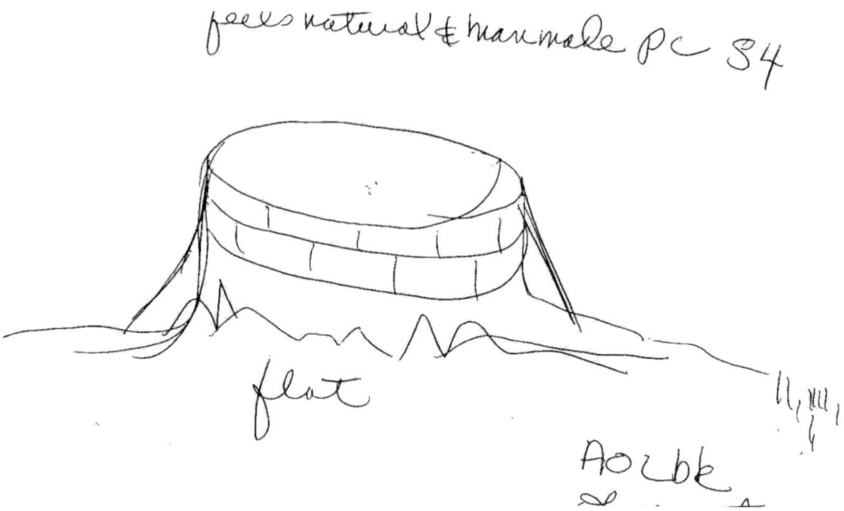

Remote viewing-produced sketch of amphitheater in El Jem, Tunisia compared to photo of the actual site. (Courtesy, Uzita Travel)

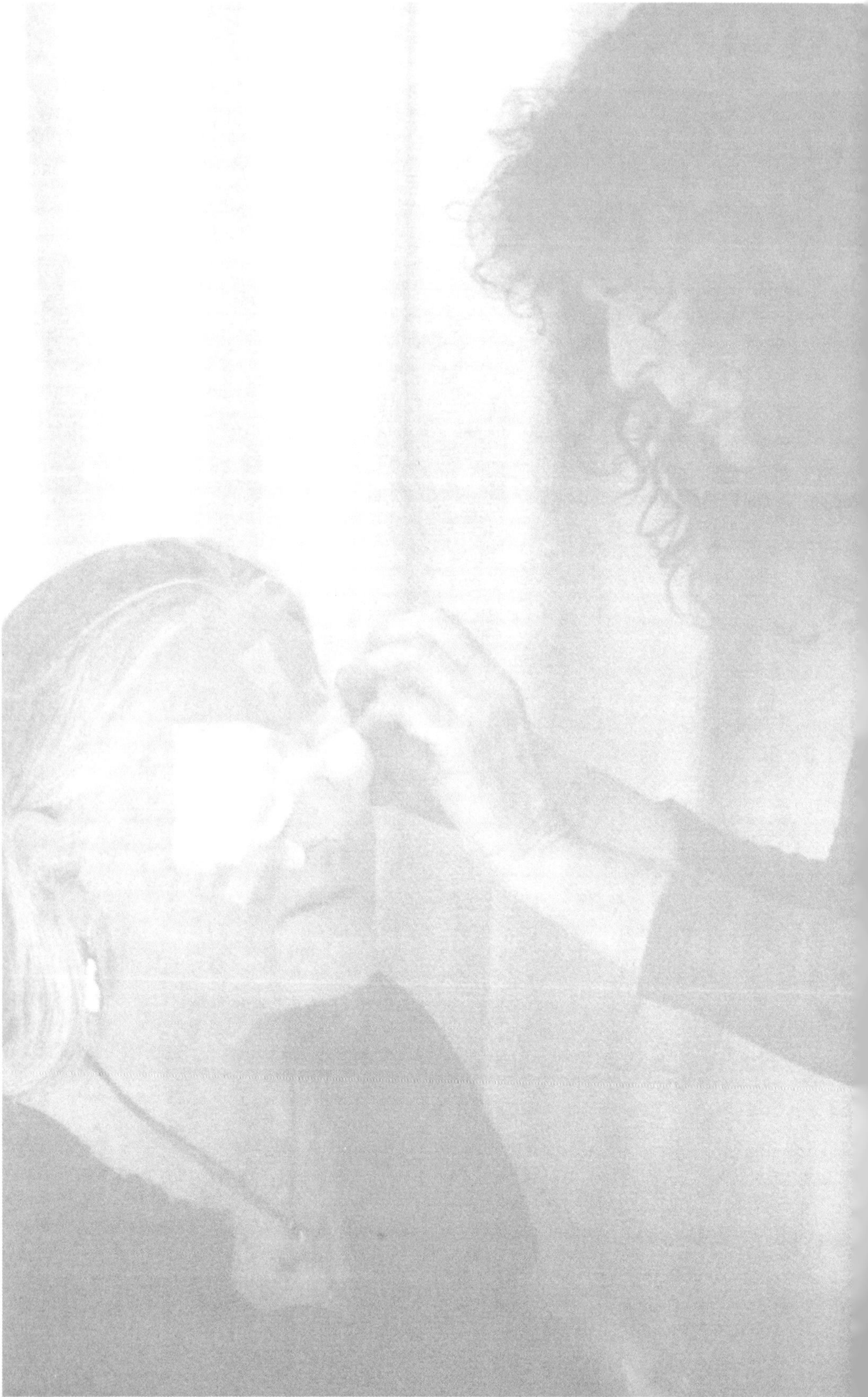

Chapter 14
Remote Perception

Remote viewing might more accurately have been labeled "remote perception." The process is much more than just visual, and a remote viewer does more than merely view the target. The full range of human perceptual experience—not just the visual part of it—is available to a remote viewer, but the term "remote viewing" has a history and is so widely used now, it seems likely to stick.

However, it doesn't hurt to think of remote viewing as one of the members of the remote perception family. In this chapter, I will describe some of the other members of this family. There are three of these that I think are most important to know about. They are presentiment, the Ganzfeld protocol (often referred to as simply "Ganzfeld," or "the Ganzfeld"), and precognitive remote perception (PRP). I have listed them here in reverse order of similarity to remote viewing. Presentiment is the least similar, while PRP is the most like remote viewing.

Presentiment Experiments

The presentiment experiment design was created by Dr. Dean Radin, a research psychologist and one of the most respected parapsychologists in the field. Have you ever had a sudden, unexpected sense of dread, a hunch, or a feeling that something bad was about to happen, but you didn't know what? Somehow you anticipated that something bad was soon to occur, and then it did. Beginning in the mid-1990s, Radin started thinking about the many stories throughout history where people "had a bad feeling" before something traumatic occurred, even though, according to the usual rules of the physics, they could not possibly have known it was going to happen. This is called a presentiment, or "pre-sentiment." Radin wondered if these premonitions were real, and if he could develop an experiment that would detect them.

Presentiments are not conscious, fact-based things. One does not consciously know what is going to happen based on some obvious warning. One doesn't even really know that something bad is about to happen—only that one is experiencing an unexplained worrisome feeling of anticipation or dread of something about the future. This is why I don't consider presentiment to be precognition, which is conscious awareness or knowledge about a future event. Presentiment seems to be a primordial reaction and, as shown by the experiment Radin devised, a person doesn't even have to be aware they are having it. But our subconscious somehow recognizes it, and our autonomic nervous system—the part of our body that governs our heart beat, blood pressure, sweat glands, and involuntary processes such as digestion and liver function—definitely recognizes something is about to happen, and it experiences this up to several seconds (and perhaps longer) before the event actually occurs.

Radin's presentiment experimental model goes something like this: a research participant sits in front of a computer and is hooked up to one or more body monitoring systems—blood volume, heart rate or, as in the earliest experiments starting in 1997, skin conductivity monitors. The participant then watches the computer monitor. On the computer's hard drive is a pool of photos. Some of these photos have calm or pleasing content (such as landscapes, babies, puppies, and the like) and others have disturbing content (gruesome shots of accidents or autopsies, close-ups of spiders or snakes, and so on). A program in the computer randomly selects between either a calm or a disturbing photo, and after a brief pause displays the photo on the screen for a few seconds.

When a disturbing photo becomes visible on the screen, the participant's skin conductance (or whichever autonomic function is being monitored) changes rapidly and dramatically. This is normal. Your body will respond that way after you unexpectedly encounter a shocking scene.

But what is interesting is what happens in the few seconds *before* a disturbing photo is displayed. If the photo is of a calming scene, the body remains calm before the photo is presented, just as we would expect. But if the photo about to be revealed is one of the shocking photos, it is as if the person's subconscious "sees" a horrible image coming in the near future, and prepares the body to react. The skin conductance or other bodily measure begins to ramp up its activity—not only before the photo appears on the computer monitor, but in some experiments before the computer has even selected the photo. The body's underlying fight-or-flight mechanism begins to kick in, even though no terrible photo is yet visible. According to the currently-accepted laws of science—in this case, even those of quantum physics—this is impossible. Yet it fits right in with our discussion of remote perception.

Presentiment experiments involve randomly showing calming or disturbing photos to a research subject and measuring whether there is an autonomic nervous system arousal in advance of the disturbing photos being exhibited.

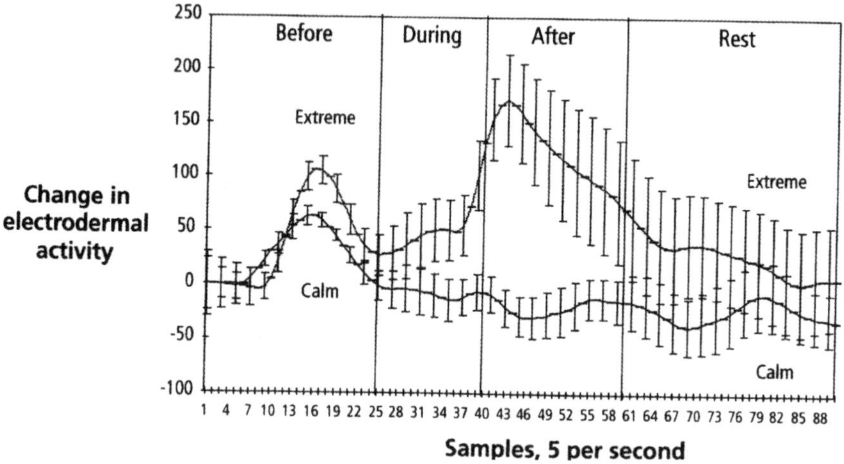

This graph charts the average change in skin conductivity (an autonomic response) over several experimental trials of one of many presentiment study participants. Note the differences in the peak response between the "calm" photos and the "extreme" (disturbing) photos that were displayed in the "before" time epoch. (Courtesy, Dean Radin)

The presentiment effect is not an isolated occurrence documented once by a single researcher. Since Radin reported the first presentiment experiments in 1997, there have been many replications performed by a number of different researchers in different laboratories. Some of these researchers have experimented with other kinds of stimuli; one used randomly-selected pleasing tones versus loud, startling tones, for example. Different physical measures have been tested in addition to skin conductance, such as heart rate or reaction of the pupils of a participant's eyes (which change suddenly in advance of a disturbing stimulus, but not a calm one).

Researchers have even discovered a presentiment effect in other perceptual experiments that had nothing intentionally to do with parapsychology research. Because the original scientists in those experiments had not even thought that a presentiment effect existed, they had never screened their data for it. But parapsychologists who knew what to look for found three separate studies (studying people with animal phobias, people engaged in risky gambling behaviors, and exploring emotional priming in language perception) that showed presentiment effects occurring just before the subjects in the experiments were presented with disturbing or unsettling stimuli which is right where one would expect to see it if the presentiment effect is real.

Why do I include presentiment under the umbrella of "remote perception," considering that the participants in these studies usually have no conscious awareness that they have reacted to a stimulus? If the participant's body reacts before seeing disturbing scenes, yet does not react in advance of calm ones, then something in his or her brain or mind must be recognizing the significance of what is about to occur. This suggests that somewhere outside of conscious awareness — in the subconscious, if you will—there is perception going on. The subconscious must be detecting in advance the kind of sensory input that is coming and must process the input and recognize what it means. The natural bodily reaction ensues in anticipation of what it is about to see.

From this perspective, it would seem that presentiment is indeed a remote perception process, even if the perceiver is not consciously aware of it.

Ganzfeld Experiments

If you weren't already familiar with it, a Ganzfeld experiment might look strange to you. In one room, you would find a person, called a "receiver," lying comfortably in a recliner, headphones covering his or her ears, with half of a ping-pong ball taped over each eye, basking in rose-colored light. In another room (or even another building altogether), you would find the "sender," a person concentrating on an image or video clip, trying to get the receiver to psychically perceive the same image or video that the sender is concentrating on.

"Ganzfeld" is German for "whole field." The concept behind it has been known for centuries. If a person's vision is restricted to a blank canvas, he or she will eventually start to hallucinate as the brain strives to find some kind of visual landmark to focus on. Taping halved ping pong balls over the subject's eyes and bathing them in rose-colored light creates just such a uniform visual field. The headphones play pink noise (static—like white noise with the more piercing high-pitched sounds reduced) into the receiver's ears to contribute to the uniform sensory field. The room itself is fitted with a heavy, sound-proof door, shielded as much as possible against sound and electronic transmissions.

The image or video that the sender is looking at is randomly selected from a target pool of four possibilities. The sender's job is to attempt to mentally transmit the visual impressions back to the receiver. Both receiver and sender are secured in their respective isolation chambers behind closed doors throughout the 30-minute experiment. These security measures prevent them from communicating in any other way besides ESP.

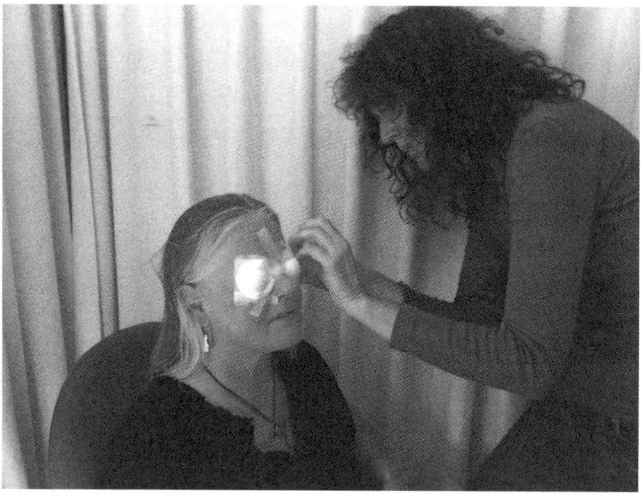

A Ganzfeld participant being prepped for an experiment. (Courtesy, Dean Radin)

Ganzfeld experiments establish a mild altered state that, the theory goes, allows the receiver to be more sensitive to ESP signals passed along from the mind of the sender. While the sender focuses on the picture, the receiver describes impressions and imagery that flow through his or her mind. The assumption behind this experiment is that the sender communicates information and impressions via some kind of mind-to-mind connection—in other words, telepathy. I will shortly explain why this assumption may not be accurate, and that remote perception might be a better explanation for what is going on in a Ganzfeld experiment.

Throughout the many Ganzfeld experiments over the years there have been many instances where the receiver's description matched the target picture. However, the real measure of success lies in the statistical analysis. After the sending phase wraps up, the receiver is kept isolated, and shown the four targets from the original pool, then asked to pick from among them the target that the sender had been focusing on. In a situation like this, mere guessing would give you a success rate of 25 percent. But if ESP were really working in Ganzfeld experiments, you would expect a consistently higher success rate.

As it turned out, this is exactly what occurred. After thousands of experimental trials involving 3,650 participants in 108 experiments performed over four decades in a number of different laboratories around the world, the cumulative accuracy rate for Ganzfeld experiments is in the neighborhood of 32 percent, yielding a highly statistically-significant result.[2]

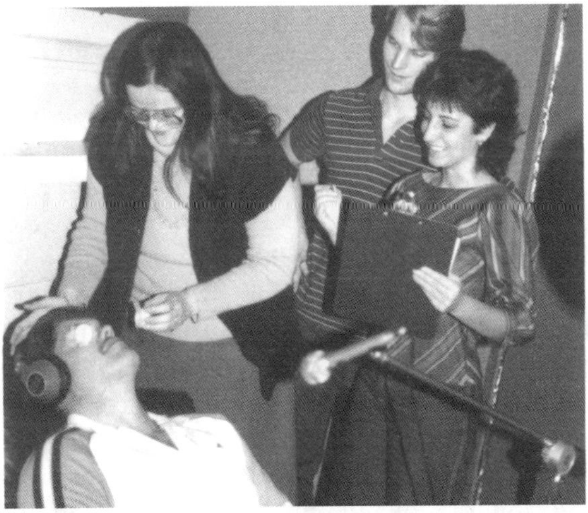

An early Ganzfeld experiment.

The Ganzfeld process was envisioned as an experiment in telepathy. I suspect, though, that rather than involving telepathy Ganzfeld experiments might just as likely, or maybe even more likely be a kind of remote perception that resembles remote viewing more than it does telepathy.

Think of it this way—the receiver in a Ganzfeld experiment serves much the same role as does a remote viewer in an RV session as he or she attempts to perceive the remote viewing target. The picture

2. Critics and researchers have heatedly debated the accuracy of the statistics showing a positive effect in the Ganzfeld experiments. But the evidence is mounting that the reported results are indeed correct.

or video clip as used in the Ganzfeld setting would be the remote viewing target, while the sender in the Ganzfeld would merely fill the role of the "beacon" team in a typical outbounder remote viewing protocol, as described in Chapter 9. So maybe what is really going on in a Ganzfeld experiment is that the "receiver" is doing the equivalent of remote viewing the location of the beacon person in order to home in on the target, which in this case would be the picture or video clip target. If this is the case, then rather than relying on telepathic communication, a Ganzfeld experiment would actually just be another form of remote viewing experiment. Thus Ganzfeld experiments would actually be a form of remote perception.[3]

PRP, or "Precognitive Remote Perception"

When the Puthoff and Targ remote viewing experiments first debuted, they made a huge splash in the world of scientific parapsychology. For years ESP research consisted mainly of tedious repetitions of card-guessing, dice-rolling, or similar experiments. But remote viewing was a protocol that was interesting and productive, and yielded tangible, visible results. Success wasn't just demonstrated by statistically analyzing the numbers from dice and card experiments. Statistical evaluation of remote viewing was still important, but now there were concrete, observable results that could be shown to anyone as evidence that ESP had worked.

It was not long before researchers tried to replicate the Puthoff/Targ experiments. Some of these were reasonably faithful to the protocols that the remote viewing experiments followed, and generally produced good results. But others strayed too far from the formula, changing crucial elements and, as a result, had much less success.

3. Some people have objected that maybe, instead, a remote viewing "outbounder" experiment is not remote perception but instead telepathy, where the beacon team is actually passing information to the viewer mind to mind, rather than the viewer merely homing in on the site and perceiving it directly. However, there are a number of factors limiting the telepathy explanation, among them the fact that the viewer often detects details and features of the target which are inaccessible to or unobservable by the beacon team, such as the tops of buildings or internal features of objects or structures at the target. Add to this that eliminating beacon teams and instead using coordinates or other ways of pointing a remote viewer at a target work quite well, in spite of telepathy not being a factor in those cases.

One who got it right was a then-graduate student named Brenda Dunne, who in the mid-1970s came across Puthoff and Targ's remote viewing work. With the blessings of her faculty advisor, she began her own attempts to duplicate the SRI results with her research partner, John Bisaha. Dunne and Bisaha decided to do things a little differently, however. Though they copied the basic SRI remote viewing approach, they decided to try a variation of it described by Targ and Puthoff in their 1976 paper. This variation involved doing the remote viewing before the target was chosen and before the outbound beacon team went to the target.

It went like this: They created a pool of 100 targets located within 20 minutes of the research facility where the viewer was isolated. At the appointed time the perceiver would perform a session attempting to describe where the beacon team would be as much as a day later. The results from this session would be locked away where no one involved in the "outbounder" part of the experiment would have access to them, and the perceiver would be excluded from any contact with anyone involved in that stage of the project as well.

In the earliest PRP experiments, before the viewer began the session the beacon person or team would have headed out with a randomly selected subset of 10 targets from the pool, concealed in envelopes. The beacon would drive around aimlessly until the session was set to be over, then blindly and randomly select a target from among the 10 that had been brought along. He or she would then go to the location identified and remain there for a specified time, just as the standard SRI outbounder remote viewing protocol required. Later, a judge (or in some cases, a set of judges) would be given the information for all 10 targets from which the actual target had been drawn, and compare those to the results produced by the viewer. The judge's goal was to use the viewer's results to try to determine which of the 10 locations the beacon had actually visited. Dunne and Bisaha repeated versions of this experiment several times, with considerable success. (In later PRP experiments, this procedure was still generally followed, but the target might be selected as much as a day, rather than a few minutes or hours later, and some of the targets were as far away as the other side of the world.)

Dunne eventually moved on to Princeton University, where her research caught the attention of a prominent physics and engineering professor, Robert Jahn, who was dean of the university's

School of Engineering and Applied Sciences. Together they formed the Princeton Engineering Anomalies Research (PEAR) laboratory and engaged in a range of parapsychology studies delving into the nature of consciousness. Dunne introduced Jahn to her remote viewing replication methodology, and they quickly realized that remote perception was a better—and more correct name—for what they were doing. They added the word 'precognitive' to reflect the future-time element of the protocol, and "precognitive remote perception" was born. Boiled down to its essentials, PRP is a remote viewing session done against a target that is always in the future at the time of the viewing. Thought of in another way,

Dr. Robert Jahn and Brenda Dunne, co-founders of the Princeton Engineering Anomalies Research Laboratory. The PEAR Lab conducted a large body of significant remote perception research.

the target in PRP is remote in two senses of the word: remote in space, and remote in time (similar to associative remote viewing).

Making the remote viewing protocol precognitive eliminated any chance for the perceiver to gain information about the target in any but an extra-sensory way. This made it more difficult for critics who questioned the work to be taken seriously if they tried to say that accurate data must have been produced by sloppy experimental protocols.

And there were indeed successful results. As I wrote in Chapter 12, the PRP results over more than two decades of experiments yielded a highly significant effect at p = 3X10-8. If you remember what you learned from that chapter, this is the same as p=0.0000000003, or one chance in 300 million that such results could be accidental.

There are other, less rigorous ways of doing remote perception. "Scrying" is one approach where a perceiver uses a crystal ball, or a half-smoked mirror, or looks into a dimly lit, uniformly colored space to invite a visionary experience (perhaps to establish a uniform or restricted visual field similar to the Ganzfeld experiment). Psychometry is yet another form of remote perception. It involves the perceiver holding an object in his or her hand and attempting to perceive the object's history and past associations.

As far as I have been able to tell, neither of these two older approaches to remote perception, nor any others like them have been carefully enough researched to get a clear picture of how effective they are or how closely they may resemble some other form of remote perception that has been more fully studied.

All that means is that there are plenty of unknowns in the field for science to explore, if scientists have the courage to open their minds to the possibilities.

Chapter 15
How Do I Learn Remote Viewing?

Ingo Swann had a reputation as a hard task-master. My fellow soon-to-be remote viewing students and I heard over and over that he expected anyone learning remote viewing to take it as seriously as he did. And if you didn't, there would be no mercy shown nor quarter given.

That was almost all of what we knew about Ingo's personality when we showed up at the huge SRI-International corporate campus in Menlo Park, California in January 1984. On this, our first remote viewing training session, we were going to be under his thumb for the next two weeks, and we weren't sure what to expect. Should we look forward to it...or dread it? We also didn't know what to expect from ourselves. Would we even be able to learn something that mainstream science claimed was impossible? Did we have what it took to master remote viewing well enough to be useful to the Army and our nation?

Then, on the bright and sunny morning of January 16, 1984, it was time to shake hands with the legend. Our worries turned out to be unfounded. Ingo could indeed be gruff and stern. But he was just as often gracious, fun, and fascinating, which is what he was that morning at our first meeting. That very afternoon the training began, and it was engrossing from beginning to end. A mixture of classroom academics and hands-on exercises, Ingo Swann's methods involved lectures, note-taking, essay writing, practical exercises, and many, many remote viewing sessions supervised (or "monitored") each and every one by Swann himself.

We went through the next twelve months alternating between two training venues. The first was the place where we began our training, in Menlo Park, California. Here we met in a windowless, secure suite of rooms built like frosting on a cake onto the roof of SRI's Radio Physics Laboratory building. The other location was in New York City, at a similar facility in an SRI-leased building in Midtown Manhattan.

In either place, the routine was pretty much the same. It was like being in school again, only on steroids—learning something far more unusual and exciting than anything I had studied in college. One of the most important things we learned was that remote viewing really could be taught. That in itself was quite a revelation. In the ensuing years we confirmed over and over again for ourselves that what we learned really was useful to the Army and the country.

Ingo Swann at a 1985 exhibit of his masterpiece "Millennium," dressed much as I remember him from our first meeting at the start of our remote viewing training. (Photo by Jeannie Betters)

As you have read this book thus far, the odds are that you have already asked yourself these two questions: Can I learn to be a remote viewer? And, if so, how? Yet not everyone I tell about remote viewing truly wants to know. Some people are afraid of ESP in general, and therefore squeamish about remote viewing. Some simply don't believe in it. Others find remote viewing mildly interesting, but not interesting enough to make room for in their busy lives. But every now and then some adventurous individual feels excited, motivated, intrigued, or even driven to learn this amazing skill. Perhaps you will turn out to be one of them. So how does a person learn remote viewing?

There are three ways:

- By the seat of your pants.
- By how-to media.
- By live training from an experienced teacher.

I'll consider each of these in turn.

By the Seat of Your Pants

The phrase "by the seat of your pants" comes from the early days of flying, when brave men and women climbed into contraptions with wings made of wood, wire, and cloth, and took off into the sky with little instruction and trial-and-error as their only textbook. They either crashed and died, or they learned to fly.

Like many aspiring remote viewers, early aviators learned by "the seat of their pants." (Library of Congress photo)

Some people try to learn remote viewing the same way. After all, remote viewing is based on an innate human ability, so why can't you develop the skill with just basic reasoning and a little practice? The do-it-yourself method goes something like this: Look around for what you can find in books and online. For example, you can go to my website and find the original controlled remote viewing manual that I and my colleagues wrote for the Star Gate program back in 1986 (it is located at http://rviewer.com/crvmanual/index.html). You can also go to books by other former Star Gate members or people who have been active in the field since the beginning such as Stephan Schwartz or Russell Targ. On your own you can simply take all the pieces and try to put them together in a sensible way that produces results.

Sounds simple, doesn't it? Well, maybe not. I have met many remote viewers and aspiring remote viewers. Occasionally, I come across people who have managed to make headway and develop some skill all on their own. One of these, Daz Smith, is an example of someone who has turned out to be a proficient remote viewer while being mostly self-taught. I am sure there are a few remote viewing savants out there who remain anonymous to me, but by far most of the ones I know who have tried seat-of-the-pants have found the approach both frustrating and generally unsuccessful. I am not saying it can't be done. I am not saying you shouldn't try it. But don't doubt either remote viewing or your own potential just because you have trouble learning it that way.

Sometimes the "seat of the pants" learning approach taught some hard lessons. (City of Toronto Archives)

In Chapters 5 and 6, I talked about how you could set up your own easy remote viewing experiments. I also gave you an idea of how to do a simple remote viewing session. If you want to try to learn how to do it on your own, that is a good place to start.

How-To Media (Home-study Courses)

A different approach to learning remote viewing involves video or internet-based training media. Video remote viewing training was pioneered by Ed Dames and his associates early in the public remote viewing saga. Ed's video course consisted of a stack of VHS tapes which instructed students in his version of remote viewing, which he called "TRV" (for "technical" remote viewing). Others have since joined the field and upgraded to DVD format or even subscriber-based e-learning offered online. Some students have made good progress with video training, but others have not.

A video may be effective in teaching you how to change a faucet washer. But imagine you were trying to learn the violin using only a book or some video lessons. You could succeed in develop the skills required for the basics. But how many people are able to develop the more complex skills needed to actually play the violin skillfully using only books or video tapes? There are some; but others who try it eventually become discouraged. To be sure, a book or videotape can help you get started, and there is nothing wrong with course offerings that do that. But not all are of equal quality. Some take a simplistic approach in order to make the product easily marketable. That process of dumbing-down leaves out many principles important to successful remote viewing. Other courses are hawked by people with only pretended expertise in remote viewing, and don't teach real remote viewing at all. (I will shortly offer some guidelines to help sort this out.)

I have not traditionally been an advocate of video-based or other forms of home study for those who really want to learn remote viewing. But not everyone has the resources or ability to take advantage of live training from a competent trainer. Yet, like anyone else these people deserve a chance to learn the skill, and for them home-study may be the only option. I feel strongly enough about improving the quality of such home-study courses that I have collaborated with an experienced video-training production company to create a home-study remote perception course that I truly believe will enhance the learning experience of those for whom live training is, for whatever reason, difficult or unavailable. You can find my course at http://www.remoteperception.com

Live Instruction[1]

If you have the resources, you should learn remote viewing the best way possible, which is a live training course from a qualified, competent instructor. But you may be confused by the numerous commercial schools and programs you have heard claiming to be able to teach you how to become a star remote viewer.

You are probably wondering which of these are legitimate (or whether any of them are) and how to tell the difference between those that are trustworthy and those that are of dubious quality. The fact is, there is a lot of difference in quality from one program to another, and there are some that even verge on being fraudulent. I will not make any recommendations as to which program to try, not even my own, since your individual needs may affect which one is best for you. You will just have to do your homework. But I will give you some suggestions on what to look for in a legitimate program, and follow up with some guidance on what to avoid.

A top-of-the-line remote viewing class will have three main elements: 1) It will teach you the essential theory—the "why" behind remote viewing; 2) It will teach you the process—the"how" of doing remote viewing; 3) And while you are in the class it will give you the opportunity, under supervision of an experienced instructor, to do a number of blind remote viewing sessions yourself to make sure you know how you are supposed to do remote viewing, and give you some initial experience on which to build with further practice. I will cover each of these in turn.

Theory Instruction

It was Hal Puthoff's and Ingo Swann's philosophy when they were instructing us that understanding the theory was essential. Merely being given a "recipe"—a step-by-step instruction on how to do

---

1. Those already familiar with how I approach remote viewing training will recognize that the material in this chapter reflects my own teaching philosophy and the way I structure my courses. I owe my point-of-view to my original training with Swann and Puthoff, reinforced by my experience teaching remote viewing starting in 1984 and lasting to the present. Inevitably, my judgment for what makes a good remote viewing program is influenced by my background. But it has worked well for my sudents for more than thirty years.

remote viewing (which is what some of the shorter, less expensive, or even free courses offer) was not nearly enough. To Swann and Puthoff, if you didn't know how and why the remote viewing process worked—at least as much as is presently known—then you would have a hard time understanding why you were supposed to do certain things to make it work. And you might also be tempted to stray from the tried and true approach to substitute something else that sounded good but wasn't as effective.

This tendency to introduce new practices or change well-established ones to something else just because it "seems to make sense" is one big reason for the variations among the descendants of the original CRV methodology. Many of these changes are unnecessary or even counterproductive to success. Without understanding the theory and principles it is often no longer clear why some idea for changing the system or process would actually not be an improvement. I endorse Puthoff's and Swann's principle even today, having seen too many times where performance suffered due to lack of background understanding.

Finding a remote viewing course with a good theory component is difficult. Many self-anointed instructors know little or nothing about the theory. They invent their own to fill the gaps, or leave it out altogether. Even some of the veteran instructors don't recognize why theory is important, or didn't themselves retain their earlier understandings very well, and thus they short-change the principle. Some of the third-generation instructors either learned no theory, or learned it incorrectly from their own instructors and thus perpetuate errors. Some who do know the theory leave it out so their courses can be shorter and they can process more paying customers. Unless you just want to take a short course to satisfy your curiosity (and that's all many people are interested in), you should make sure there is theory instruction in any class you take.

Learning How to "Do" Remote Viewing

This is the part of the process people usually think of when they talk about learning remote viewing. But proper remote viewing training should involve much more than memorizing a step-by-step "recipe" for carrying out the remote viewing process. As I discussed above, learning the theory is where it all starts. Then one must learn the technique, the actual process or sequence that one practices. This should include more than just the individual steps one takes to remote view. The student should also be guided to understand what to expect at each step, and how to recognize and deal with the various phenomena, mistakes, and behaviors that are associated with the process, such as analytical overlay, sketching procedures, ideogram formation (in CRV and its derivatives), internal editing, aesthetic impact, and so on.

Likewise, learning the remote viewing process isn't just about being told how to do it, then simply doing it. A quality remote viewing course will provide appropriate drills, practical exercises, and review exercises to help you master the skills involved. Practical exercises simulate different parts of the remote viewing process to help students gain a preliminary idea of how a certain part of the skill is performed or what a certain part of the experience is like. Then when the student performs an actual remote viewing session, each part of the process will be more familiar. The learner will be able to master the skill more quickly and with less effort, involving less trial and error. These drills and exercises will vary depending on the type of remote viewing being learned, but in nearly all cases they should serve, among other things, to help the viewer recognize and learn to properly report sensory and cognitive experiences, perform complex actions, and so on.

Exercises may involve quizzes, writing review essays, or challenge-and-response drills. Recent psychological research shows how important these exercises are for both factual learning, and skill-development and retention.

In-Class Remote Viewing Sessions

If you've ever tried to learn how to play a musical instrument, or ride a bike, or swim you know that merely watching while the instructor plays an instrument, rides a bike past you, or swims the length of the pool is never enough to learn how to master a skill. You have to do it yourself, many times. In the same way, merely having a remote viewing instructor demonstrate a remote viewing session for you won't teach you remote viewing. But that is what you might encounter in some of the less-helpful courses. You need to have the opportunity to try it yourself, and not just once or a small number of times.

The author with his own remote viewing instructor, the late Ingo Douglas Swann (1933-2013).

As you might expect, there is more than one approach to trying it yourself in a remote viewing course. Many live-instruction courses—particularly those that teach large numbers of people at once in the same room—merely have each student work individually on a practice session or two while everyone else in the room is doing the same thing. This gives you the least amount of individual attention, and hence the least amount of guidance and correction from the expert. The learning quality is diminished, but on the positive side these sorts of classes usually cost much less per student, and require a shorter time commitment. They can be an adequate introduction or just to satisfy your curiosity about remote viewing.

Another approach is to have the instructor explain the process, then pair up students to work with each other. This is an improvement over the group approach, but the downside is that the two inexperienced students may reinforce each others' mistakes and fail to correct each others' technique. Having an instructor explain and demonstrate a skill for you, and then sending you off to practice it on your own or with another beginner is not the best approach to learning remote viewing.

By far the most successful approach is to teach the way Ingo Swann taught me and my colleagues - one-on-one remote viewing sessions between each student and the remote viewing instructor.[2] What this means is that each student gets the benefit of having an experienced instructor who already knows the correct practices, the mistakes and pitfalls, and how best to guide a student through the remote viewing learning process in every practice session.

I find that this allows me as the instructor to quickly identify when a student has taken an unproductive tangent. I can also recognize the individual problems students may develop and offer strategies to correct them. Further, I can immediately help students adopt better techniques when dealing with some of the subtle perceptual experiences one must master to become a competent remote viewer. None of this is available in a course where the student only does individual sessions or where students work only with each other, or where the instructor only works one or two times with each student.

Follow-up programs after the in-house part of the course is over

If you want to become proficient at remote viewing, your learning can't stop with the end of a formal class. Just like any other complex human skill, you have to reinforce the skills you've learned, and cement and expand them through further practice. You will find that it is helpful to have access to at least some mentoring or continuing education to guide you after your course is completed. Thus, the optimum remote viewing course should offer some at-home follow-up, either as part of the course structure or as an additional offering. This may take the form of individualized support, or some kind of online community-based access. The availability of this kind of support, and how individualized it is, can be a significant indicator

---

2. As an example, the standard in my classes is to keep the student-teacher ratio low - ideally two students per instructor. If I have more than two students I usually hire an additional instructor to assist. If I have six students, I hire a third instructor; but that is as large as I allow a full-featured remote viewing class to get.

of course quality, and should be taken into account when making a decision about what training direction you should pursue.

What you don't need

It doesn't hurt to mention briefly what you don't need in a remote viewing course. You will notice I didn't say anything about learning the history of remote viewing. It's not that I don't think remote viewing history is important—it is, and everyone interested in remote viewing should have a fundamental, accurate understanding of its origins. But knowing the history isn't necessary to learn remote viewing. There is plenty to learn in a live class without going through the history, which often degenerates into "war stories," many of them inaccurate or self-serving on the part of the instructor. An exception to this is when a story can highlight a problem or a solution to a problem, or illustrate some important point in the actual learning or using of remote viewing. In those cases, a little history can be essential.

You also don't need an emphasis on what I call "anomaly" targets—UFOs, mysterious creatures, hauntings, religious events, ancient mysteries, one's future life course, and so on. In other words, targets that are either impossible or unlikely to give you feedback. Why is this? If your goal is to learn remote viewing, you need to work with targets that have a real-world, verifiable background. If you cannot be presented detailed feedback for a target you have remote viewed, then you can't tell whether you got it right or not. Maybe there is entertainment, but there is no learning value in the time you (or your instructor) might be tempted to spend trying to remote view an anomaly or mystery target.

When you have a choice between several skilled instructors who offer quality curriculum, what makes the difference between a good class and an excellent one? The answer is in the number of hours spent in class, the number of sessions done with direct instructor participation, whether that participation is individualized, and how many hands-on practical exercises are included. A course with more of these features will also be more expensive, for two reasons: first, more of the instructor's time and energy will be required per student, and second, there will be fewer students in the class to share the cost of putting it on.

Which course and instructor you choose depends on your goal. If you are serious about mastering remote viewing, you will want higher-quality, more intense instruction, and a lower student-teacher ratio. If, on the other hand, your goal is largely just to satisfy your curiosity about remote viewing (and there is nothing wrong with that), or to get a basic introduction at a starting point, you may want to go with a shorter, less intense, and less expensive course with fewer student-worked sessions, all or most of which are done in a group setting.

What to Beware Of

In the United States, the medical profession has the American Medical Association to make sure people who provide medical services at least have a firm foundation in the skills they claim to provide. There are organizations that certify that those who teach at medical schools meet minimum qualifications. The same is true of other long-established professions. However, remote viewing is a relatively new field that does not yet have that kind of oversight.

Starting in late 1995 when the government remote viewing program first became public knowledge, there have been people who have encroached on the field, claiming falsely that they were former Star Gate remote viewers or remote viewers from "overseas intelligence services" (though which overseas services are never disclosed) and who say they can teach you remote viewing.

There are others who hang out their remote-viewing shingle who sincerely believe they know what remote viewing is, how to do it, and how to teach it. But, to borrow some infamous words from former Secretary of Defense Donald Rumsfeld, these folks unfortunately "don't know what they don't know," and often what they teach is mistaken or incomplete.

There are even those who do understand remote viewing quite well, but choose to "phone in" their teaching performance, taking short-cuts, or oversimplifying to the point of depriving their students of a sound learning experience. It can sometimes be hard to distinguish a less-competent or lazy instructor in advance of paying for a course. But there are some guidelines that can help you navigate through the variety of offerings. Here are just some things to be wary of:

Claims about "brainwaves" or needing to be in some kind of brainwave state (as measured by an EEG machine), such as 'alpha," "delta," or "theta," to be able to remote view. Research at SRI showed that remote viewing could be successful no matter what brain-wave state the viewer was in. If someone tells you that you must cultivate some specific brain-wave profile or another to learn remote viewing, you can be sure that they are uninformed.

Associating remote viewing with astral travel, "spirit" guides, channeling, or the like. Whatever value astral travel or spirit guides may or may not have in their own right, they have absolutely nothing to do with remote viewing. Someone who lumps remote viewing in with these doesn't understand the differences.

Claims to having learned remote viewing from some undisclosed east- or west-European (or other) country's) intelligence agency. There are people attempting to establish credibility so thoy can sell books or remote viewing training products or programs by claiming to have gotten their remote viewing skills from one source or another. But they can't reveal what the source is because it is "still secret." While the Soviets did have an ESP program during the Cold War, there is little evidence that any other country besides the United States and Russia had the kind of program that would produce competent remote viewers. Those who claim otherwise have no evidence and expect you to take them at their word. Don't trust them unless they produce real proof.

Claims to teach you remote viewing in a weekend. Like any complex human skill, becoming a successful remote viewer requires that you learn not only the concepts and theory of how to do it, but also that you practice extensively. No one can teach you everything you need to know to become an operational-level remote viewer in a weekend. Some legitimate remote viewing experts, notably Stephan Schwartz and Russell Targ, will often say that they can teach you "everything you need to know" to do remote viewing "in 15 minutes." What they mean is that they can teach you how to do a generic, basic level remote viewing experiment along the lines of what I spelled out in Chapter 5 in a relatively short period of time. This is a good starting point, but it is not the level of remote viewing expertise that many people interested in remote viewing aspire to.

Claims that you can be 100 percent accurate in remote viewing. This is just a commercial come-on. No remote viewer has ever achieved consistent 100 percent accuracy. Given the nature of both ESP and standard human perceptual psychology, such a consistent level of accuracy is not just unrealistic, it is impossible.

Claims that you can remote view even while knowing what the target is. We call this "being frontloaded," and having read this far through the book, by now you probably have a pretty good idea why a claim like this is false.

Several of these cautions are associated more with online or distance-learning remote viewing training than live instruction, but others apply across the board. Fortunately, there are instructors who are legitimate and don't make these claims to trap the unwary. But it is important to do your homework before spending your hard-earned money on what might turn out to be an unsatisfactory training program.

To wrap up this chapter, I'll give you a list of questions to ask of any live remote viewing training program that you may be contemplating paying for. I can't give you the answers to the questions. You will have to consider all the variables for your own particular circumstances, such as available time and finances, your level of interest, what your ultimate goal is, and so on. You may have to make adjustments or compromises based on that assessment. However, the answers you get to these questions will allow you to compare and contrast training programs and help you to make the best choice for what you need.

Questions to ask in evaluating a course you may want to take:

- What is the full cost for the course?
- How long does the course last?
- What are the instructor's full credentials? (The more

dubious an instructor's abilities, the more vague and incomplete he or she will be about how he or she acquired remote viewing expertise.)

• How many actual hours of instruction are there in the course? (This number will be only approximate, since different students learn at different rates; still the estimate should always be pretty close to the actual.)

• Is theory instruction included and if so to what extent?

• Are practical exercises and recall exercises included in the curriculum?

• Is there homework after class each day? (This is not something I've talked about earlier, but, depending on its content, homework can be a valuable way of cementing knowledge learned during the day and prepare students for new concepts coming their way in the next block of instruction.)

• How many sessions will I personally do during the course? (Obviously, the more there are, the better.)

• Will these be individual sessions where I work on my own ("solo"), or monitored by another student, or will they all be monitored by an instructor?

• What is the student-teacher ratio/class size? (Obviously, the closer the ratio is to one-on-one, the higher the course quality; I have found a two students-per-teacher ratio to be ideal.)

• Is there a follow-up program to help me reinforce the skills I learned during the course? If so, what does it consist of?

Gathering the answers to these questions can allow you to compare across various remote viewing training programs and help make your decision as to which course is best for you an easier and more informed one.

Chapter 16
Where do You go From Here (and how do you get there)?

"Now what?"

The answer to that question depends on what you think about what you've just read - whether you want to do something about what you've learned, or whether you're just ready to close the cover and go on to some other interest. If you are sufficiently fascinated by this surprising ability that each of us has within, just waiting to be brought out, you are ready to begin a challenging but a rewarding journey. It is a journey upon which many have embarked around the world in the years since the CIA finally revealed its remote viewing secrets.

There was a time when almost no one knew about remote viewing, whether inside or outside the government. The papers published by Puthoff and Targ in the mid-1970s started a firestorm of interest, but mostly in a relatively small circle of interested scientists and skeptics. Even when their book *Mind Reach* came out in 1977, it attracted relatively limited attention. Something changed, though, with the revelations from Ted Koppel's Nightline program in late 1995. Perhaps it was the broader reach of the media and the Internet. Or maybe instead it was the cachet and excitement generated from the fact that it had been the military that had used remote viewing. Or maybe the public had simply been prepared for the idea of remote viewing by movies and TV shows such as "The X-Files" —or all of the above. Whatever it was, the idea of remote viewing caught fire. Now tens of thousands of people in countries around the world have been introduced to remote viewing, and many of those people have learned how to do it themselves.

In case you do decide to set out on that journey, I have some resources to recommend and advice to give in this final chapter. First, the resources. This book isn't all there is. Its companion website offers features, media, exercises, and additional material to support what you have learned and encountered in these pages. The address is https://www.guidetoremoteviewing.com. Links from the front page of the site lead to a few subpages with information and materials available to everyone. But readers of the book have access to additional material and features. For example, there is a "Try it, you'll like it!" remote viewing trial and practice area. There are a handful of targets available in the open area for anyone to try. But those with access will find additional targets on which to start their remote viewing journey—or for those who might already have remote viewing experience to get more practice in.

Also, there is a collection of photos of many of the people you have met in this book, as well as remote-viewing related video and audio files either hard to find or unavailable elsewhere. These will be updated from time to time as new material is created. Often this may be content that will later be made available to the public, but it will be posted there first.

Those with access will also get to see the full remote viewing results for some of the most famous remote viewing cases from the Star Gate program's files, including the *Typhoon* submarine project that features so prominently in this book. There are also examples of other remote viewing sessions and results not accessible anywhere else. You will also find links to important remote viewing internet sites, and suggested remote viewing reading material, such as books, online articles, and so on—including links to reference materials listed in the book's notes section. This will also be updated as new material becomes available.

And it is easy to get the access I've been talking about. All one need do is go to http://www.guidetoremoteviewing.com and find the password protected area. Then enter a user name you select and the password 'Iwant2RV' (without the quotes), and you're in.

Now for the advice: Take it easy! By that, I mean don't try to do everything overnight—whether it is learning about remote viewing or, if you're trying to learn how to do it, don't expect to become an expert by the day after tomorrow. These things all take time, and you will be happier and less frustrated if you pace yourself and take breaks from it periodically.

Also, don't automatically believe everything you hear. I touched on this earlier, but it bears repeating: there are a lot of people out there saying many things about remote viewing. Some of them know what they are talking about, others only think they do and there are even a few that have an agenda and may be willing to deceive to accomplish it. Don't trust what any of us say until you have thought it through, compared it to what others have to say, and decided whether it seems plausible or not. This will save you time and aggravation in the longterm, and you will avoid many blind alleys and time-consuming detours. In the process you will likely come across some of the same arguing and disputation such as goes on in any newly emerging field. Don't let that distract you. Yes, the remote viewing community can be noisy, but it's worth putting up with that to find the nuggets and core experiences that lie beyond.

Finally, have fun! Remote viewing should be an empowering, exciting activity that brings you new insights both about the world and about yourself. I am convinced that because of my remote viewing experience I am a different and I believe better person than I might otherwise have turned out to be. You may find that to be true for you as well. Being good at remote viewing requires us to learn how to be absolutely honest with ourselves; and not just honest in admitting where we might be weak, inadequate, or self-deceptive, but also in admitting and accepting our strengths and positive qualities. Being successful at remote viewing means learning to acknowledge to ourselves that we are no better than we are but (of equal importance) that we are also no worse. It also means having the desire through conscientious effort to move the ratio of our positive–to–negative qualities more towards the positive side through learning, practice, and experience. Work is involved, but it is the kind of work that is rewarding, fulfilling and fun!

There is one last essential remote viewing resource left to note: the International Remote Viewing Association (http://www.irva.org) which, among many other things, sponsors the only regularly scheduled remote viewing conference. IRVA also publishes and maintains a number of information–rich websites and an enhanced newsletter (available online and print–on–demand). On the main IRVA website is to be found a bibliography listing nearly all credible published literature on remote viewing, including links to much of that material that is currently available on the Internet. Also available to IRVA members are all 90,000 pages of the declassified CIA Star Gate Archives in Adobe Acrobat PDF format, along with a helpful guide that I created and donated to IRVA to help members get started in finding their way through the many thousands of documents that make up these fascinating archives.

IRVA also sponsors an online remote viewing member discussion list for anyone from neophyte to expert, as well as the "Focal Point" remote viewing practice group, which posts a new remote viewing practice target every two to three weeks. Members of IRVA who want help polishing their remote viewing skills or want to try the seat-of-the-pants approach to self–learning will find this an excellent resource for regular remote viewing practice in addition to the targets available on the guidetoremoteviewing.com website.

The founders of IRVA gathered for a group photo in Alamogordo, New Mexico, March 1999.

Finally, IRVA members have access to the streaming video page, which contains videos of every available IRVA remote viewing conference presentation since 2000 (though access lags the most recent conference by a few years). There are dozens of fascinating video presentations available to IRVA members that nonmembers would have to purchase to watch. I also recommend to you the annual Remote Viewing Conference that IRVA sponsors, which every year brings leaders in the remote viewing field together face to face with the general public interested in remote viewing to discuss and present on the many facets of remote viewing, remote perception, and nonlocal consciousness. The conference is open to the public, but members of IRVA get a valuable discount on conference fees. The latest conference information is available online at http://www.irvaconference.org/ .

Given all that IRVA has to offer, I have negotiated an agreement with the organization to give readers of this book a free one-year introductory membership (currently a $45.00 value). Go to http://www.irva.com/student/ and fill in the block with your email address, then click 'Next.' On the following page, fill in your contact information as requested. In the "School/Instructor" box enter (once again) the following code: Iwant2RV. Then just follow the remaining instructions to start receiving your benefits! That's all there is to it. This is the same page, by the way, that I recommend to my first time remote viewing students to get their one-year free membership in IRVA. The association offers this benefit to new students of every regular remote viewing instructor.[1]

In 1987, about halfway through my time as a military remote viewer, my friend (and then boss) Army Major Bill Ray stopped me as we were going in opposite directions through the Star Gate project offices. "You know what amazes me about remote viewing, Paul?" he asked, as he chewed thoughtfully on the stem of his well worn pipe. I'm sure I looked puzzled. "What amazes me is that what we're doing here doesn't

Bill Ray and Paul H. Smith at an award ceremony in late 1985.

amaze me anymore." This from the man who often liked to explain remote viewing by merely saying "It's magic."

I understood what he meant. True, back then we psychic spies were still doing the seeming impossible. And true, we were still producing some pretty remarkable results just as you will be able to do by following the right principles and with enough practice. Back then the seeming miracles we often achieved using remote

1. The trial membership offer is not available to anyone who has been a member of IRVA in the past.

viewing didn't amaze us any more because we had been working this "magic" for so long that we were more likely to be surprised when it didn't work than when it did. And that in itself is an amazing state to be in. I hope some day soon that, just like we remote viewers in the government's former Star Gate remote viewing program, you too will find yourself just as un-amazed by some seeming miracle you have accomplished using remote viewing. Because, if you really want to, you can do it too.

APPENDIX

The chart below and the one on the next page are meant to help relieve some of the confusion and conflicting claims that have emerged since declassification of the Star Gate program. This graph shows the time in years some of the more publicly-known military personnel served in the Star Gate program. (Information obtained from official military documents.)

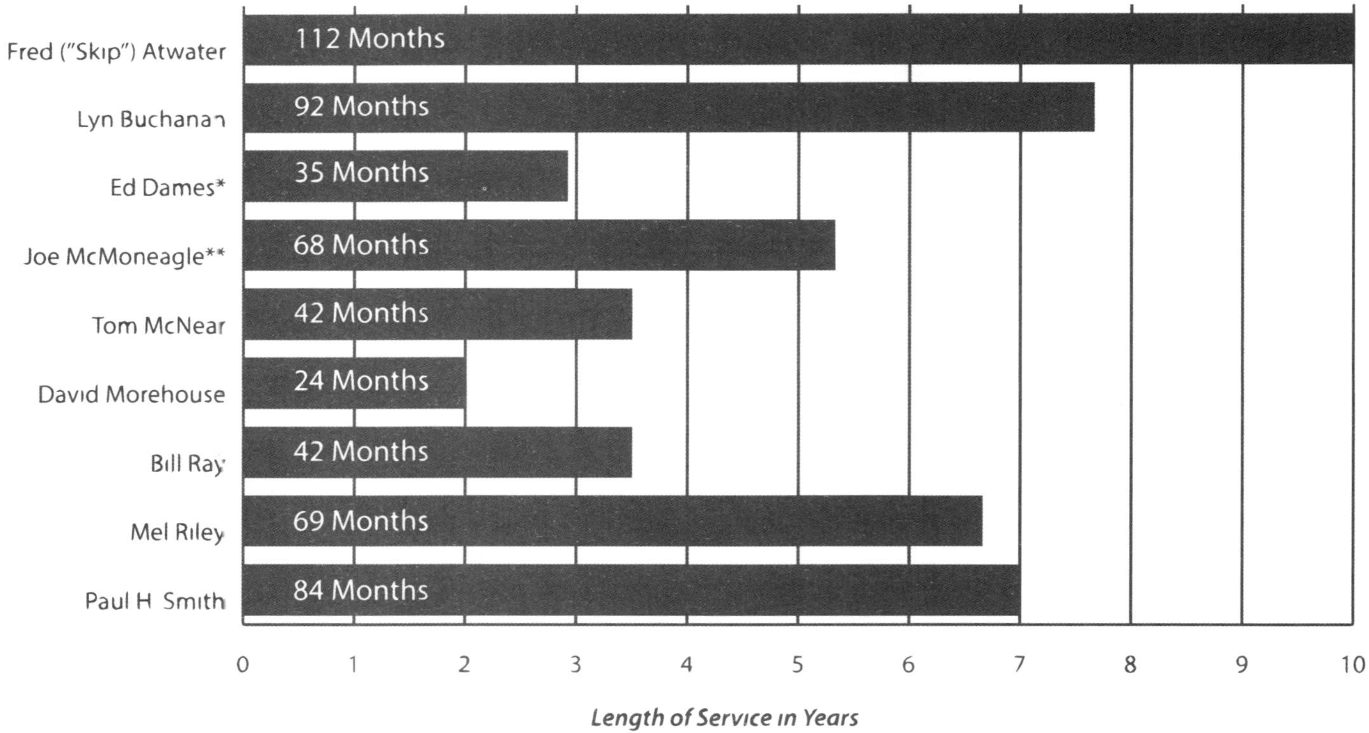

*Though not officially assigned to the organization, Ed Dames also trained for 26 weeks in 1984 with Star Gate personnel.

**Joe McMoneagle went on to serve for almost 12 more years as a research associate at SRI and later SAIC until the close of Star Gate.

This chart shows the approximate chronological start, length of service, and end point of each of the more public members of the government's Star Gate program. This will hopefully resolve some of confusion surrounding precedence and experience of the various participants. (*Dale Graff's involvement in RV began in 1975 in the earlier Air Force program.)

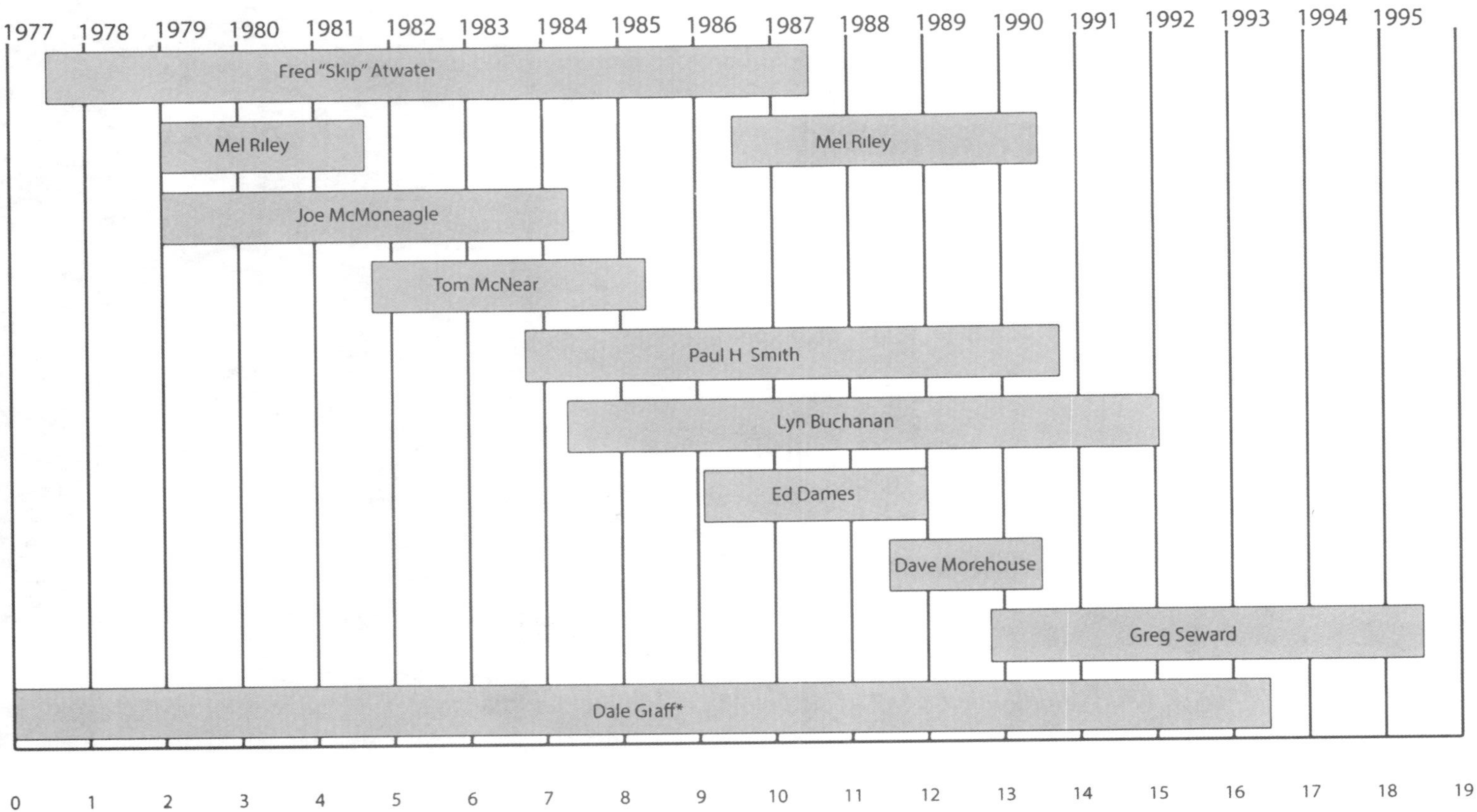

END NOTES

The following references are for those who are interested enough to want to dig more deeply into the subjects covered in each chapter. Scattered throughout are a number of the most important references on remote viewing, as well as pointers to other sources where additional information can be found.

Chapter 2 - Unleashing the Military Mind

For a more extensive account of how I ended up as a psychic spy, see Chapter 3 of my book *Reading the Enemy's Mind: Inside Star Gate - America's Psychic Espionage Program*, (Tom Dougherty, 2005).

On the history and findings of parapsychology research, including ESP, see Dean Radin's excellent book, *The Conscious Universe: The Scientific Truth of Psychic Phenomena*, (Harper, 1997). Another good account is Richard Broughton's *Parapsychology: The Controversial Science*, (Ballantine Books,1992).

Chapter 3 - The Most Paranormal Thing...

You can read Hal Puthoff's own account of how the remote viewing program got started in his article "CIA-initiated remote viewing program at Stanford Research Institute." *Journal of Scientific Exploration,* (Vol.10, No.1, 1996), 63-76. Russell Targ has written his own memoir of the program, "Remote Viewing at Stanford Research Institute in the 1970's - A Memoir." *Journal of Scientific Exploration,* (Vol.10, No.1, 1996), 367-380. Both are available on the IRVA library page linked from http://www.irva.org.

One excellent early history of the military remote viewing program is Jim Schnabel's *Remote Viewers: The Secret History of America's Psychic Spies*, (Random House, 2011). My own book, *Reading the Enemy's Mind* (mentioned above), also has an extensive historical treatment.

Dale Graff's *Tracks in the Psychic Wilderness: An Exploration of Remote Viewing, ESP, Precognitive Dreaming, and Synchronicity,* (Element, 1998), gives an introduction to Graff's important role in the military remote viewing program, and his book *River Dreams: The Case of the Missing General and Other Adventures in Psychic Research,* (Element, 2000), tells the stories of a number of the more prominent operational successes of the remote viewing projects in which Graff was involved.

Another interesting and significant book is F. Holmes "Skip" Atwater's *Captain of My Ship, Master of My Soul: Living with Guidance,* (Hampton Roads Publishing, 2001), which recounts much of Skip's involvement in the remote viewing program.

Chapter 4 - The psychics were right

Though the story of the Typhoon project in this book is the most detailed version published to date, you can find an account written by one of the people directly involved in Joe McMoneagle's *Stargate Chronicles: Memoirs of a Psychic Spy,* (Hampton Roads Publishing, 2002).

Chapter 5 - Putting Remote Viewing to Work

Jim Schnabel's *Remote Viewers: The Secret History of America's Psychic Spies,* and my own *Reading the Enemy's Mind,* give detailed accounts of how remote viewing was used for intelligence gathering.

Two books that discuss forensic ESP, including remote viewing, are Lyons, A., & Truzzi, M., *The Blue Sense: Psychic Detectives and Crime,* (MystTerious Press, 1991); and *Psychic Criminology: A Guide for Using Psychics in Investigations,* by Whitney Hibbard, Raymond Worring & Richard Brennan, (Charles C. Thomas Publisher Ltd, 2002).

A newspaper story about how remote viewer Angela Thompson Smith helped solve a murder case is Schoenmann, J., "Seeing dead people: 'Remote viewers' in Nevada help solve California murder," *The Las Vegas Sun,* (May 5, 2012). (Available online at http://www.lasvegassun.com/news/2012/may/05/seeing-dead-people-remote-viewers-nevada-help-solv/).

I will go into more detail about Kolodziejzyk's as well as the college students' successes in making money with remote viewing in a later chapter. But a good survey of the associative remote viewing research and a discussion of some of the financial aspects of it written from a scientific perspective is Bierman, D. J. & Rabeyron, T., "Can psi research sponsor itself? Simulations and results of an automated ARV-casino experiment," which was presented at the 56th Annual PA convention in Viterbo, Italy, August 8-11, 2013. (This paper can be accessed online at http://www.uniamsterdam.nl/D.J.Bierman/PUBS/2013/ARV_Casino.pdf).

George McMullen's book *One White Crow* (McMullen, G., & Emerson, J. N. (1994). Hampton Roads Publishing Company) gives examples of George's success with remote viewing in archaeology. A brief discussion of George McMullen's "Lost Colony" work can be found in Sumner, S. (2009). "Lost Colony site discovered through remote viewing." PRWeb August 18, 2009, (accessible online at http://www.prweb.com/releases/remote/viewer/prweb2759384.htm).

A detailed account of the remote viewing archaeology project led by Stephan Schwartz in the ancient city of Alexandria, Egypt is available in Schwartz, S. A., *The Alexandria Project*, (Delacorte Press/E. Friede, 1983 - and later editions). His book *Opening to the Infinite: The Art and Science of Nonlocal Awareness*, (Nemoseen, 2007; to be published in a new edition in 2016); Schwartz also has several scientific papers linked on his website (http://www.stephanaschwartz.com/category/peer-reviewed/page/10/) describing other similar archaeology projects.

Chapter 6 - A Matter of Time...

Greg Kolodziejzyk's own excellent scientific report of his extensive (and successful) associative remote viewing project, which embodied both research and investing is "Greg Kolodziejzyk's 13-Year Associative Remote Viewing Experiment results," *The Journal of Parapsychology*, (Vol. 76, No. 2 , Fall 2012), available online at (http://www.remote-viewing.com/ARVpaper.pdf).

For an accessible introduction to associative remote viewing, see Stephan A. Schwartz's article "The Origins of ARV," *Aperture* (The International Remote Viewing Association, Fall/Winter 2014, no.24, 3-9), and Chapter 12 of his book *Opening to the Infinite*.

Chapter 22 of my book, *Reading the Enemy's Mind* talks in detail about the precognitive session that described the attack on the *USS Stark*. The remote viewing of the incident was also described in a Jack Anderson column in *The Washington Post* (Jack Anderson and Jan Moller, "The Startling Vision of a 'Psychic Spy'," Jan. 9, 1997).

Chapter 7 - "Pick Your Poison" the Assorted Flavors of Remote Viewing

Many of the amazing successes of the early GRVers can be read in Targ's and Puthoff's book *Mind Reach: Scientists Look at Psychic Abilities* (first published by Delacorte in 1977, the most recent edition is Hampton Roads Publishing, 2005, which includes a new introduction and a reprint of Puthoff's 1996 paper on the beginnings of the CIA program).

The Controlled Remote Viewing Manual is available online at http://rviewer.com/crvmanual

Chapters 11-20 of my book, *Reading the Enemy's Mind* give an extensive account both of the theory behind CRV and my experiential account of the training that supported it.

For a focused treatment of Lyn Buchanan's version of controlled remote viewing, see his book *The Seventh Sense,* (Paraview, 2003).

Chapter 8 - "Can I Be a Remote Viewer?"

You can view, a documentary about remote viewing called "What We See When We Close Our Eyes," along with director E. Elias Merhige's amazed reaction to his own remote viewing success, in the bonus features of the movie on the *Suspect Zero* DVD.

In his book *Remote Viewing Secrets,* (Hampton Roads Publishing, 2000), Joe McMoneagle shares valuable insights into the experience and process of remote viewing.

Chapter 9 - Two Simple Remote Viewing Experiments You Can Do

Russell Targ's book, *Limitless Mind: A Guide to Remote Viewing and Transformation of Consciousness,* (New World Library, 2009), has a helpful section on how to start with remote viewing yourself. Stephan Schwartz's *Opening to the Infinite* also provides some useful advice.

Chapter 10 How Does Remote Viewing Work?

For a detailed discussion of experiments showing ESP in general and remote viewing/remote perception in particular can't be explained by electromagnetism see Targ, Puthoff, & May, "Direct Perception of Remote Geographical Locations," in Tart, Charles T., Puthoff, Harold E., & Targ, Russell (eds.), *Mind at Large: IEEE Symposium on the Nature of Extrasensory Perception,* (Hampton Roads Publishing, 2002); Stephan Schwartz's Opening to the Infinite; and Chapter 7 of my own *Reading the Enemy's Mind*.

Dean Radin's book *Entangled Minds: Extrasensory Experiences in a Quantum Reality,* (Paraview/Pocket Books, 2006), discusses the entanglement hypothesis for ESP.

Brian Josephson's 1991 paper, "Biological Utilisation of Quantum NonLocality," *Foundations of Physics,* (Vol.21, 197-207), written with Fotini Pallikari-Viras, gives a nuanced view of his thoughts on ESP and quantum nonlocality. The paper is available online at http://www.tcm.phy.cam.ac.uk/~bdj10/papers/bell.html

Chapter 11 - The Evidence for Remote Viewing

Two of the most important early articles on remote viewing research (and published as a direct result of Targ and Puthoff's debate with the CIA) are Targ, R.and Puthoff, H. "Information Transmission Under Conditions of Sensory Shielding." *Nature*, 252 (1974), 602-607; and Puthoff, H. E. and Targ, R. "A Perceptual Channel for Information Transfer over kilometer distances: Historical perspective and recent research." *Proceedings of the IEEE*, 64(3) (1976), 329-354. Available at (http://www.irva.org/library/articles/).

For a broader discussion of evidence for ESP, see Michael Schmicker's book *Best Evidence*, (iUniverse. 2002), which details his search for compelling evidence for the reality of ESP (though the focus is on general ESP, not specifically remote viewing).

An even more recent book with a rigorous scientific approach collects a number of leading-edge, peer-reviewed papers in one volume giving evidence of ESP is Damien Broderick & Ben Goertzel, *Evidence for Psi: Thirteen Empirical Research Reports*, (McFarland, 2015).

Anecdotal evidence gets little respect—except when there is a lot of it. When story after story builds up, each with a reasonable claim to authenticity and independent of the others, then they start to be persuasive to reasonable people. Such is the case in Sally Rhine Feather's excellent book *The Gift: The Extraordinary Experiences of*

Ordinary People, (St. Martins Press, 2005), which mines the rich archives her mother, Louisa Rhine, gathered over decades of field work. These archives consist of tens of thousands of accounts from people around the world who had significant, spontaneous ESP experiences, many of them confirmed by witnesses.

One of the best compilations of scientific research into remote viewing and remote perception is Tart, Puthoff,& Targ, *Mind at Large: IEEE Symposium on the Nature of Extrasensory Perception*, cited in the notes for Chapter 10.

An exhaustive exploration of ESP and the evidence for it is Etzel Cardeña, John Palmer, & David Marcusson-Clavert (eds), *Parapsychology: A Handbook for the 21st Century* (McFarland, 2015).

Chapter 12 - Remote Viewing: Evidence of ESP

The review mentioning the 28 remote viewing studies is most easily available as an appendix in Targ, Russell & Harary, Keith, *Mind Race: Understanding and Using Psychic Abilities*, (Villard Books, 1984).

The impressive statistics for Jahn & Dunne's precognitive remote perception experiments are available in Dunne, B. J., & Jahn, R. G., "Information and uncertainty: 25 years of remote perception research." *Journal of Scientific Exploration*, (Vol. 17, No. 2, 2003), 207-241. (Online at http://www.scientificexploration.org/journal/jse_17_2_dunne.pdf); as well as their important book Margins of Reality: *The Role of Consciousness in the Physical World*, that I mentioned earlier.

Jessica Utts's important article is "The Significance of Statistics in Mind-matter Research," Journal of Scientific Exploration, (Vol.13, No. 4, 1999), 615-638. (Online at http://www.ics.uci.edu/~jutts/JSE1999.pdf), The full military study is: Edwin C. May, Virginia V. Trask, Wanda W. Luke, Thane J. Frivold, & Beverly S. Humphrey, *Review of the Pscyhoenergetic Research Conducted at SRI International* (1973-1988), (SRI International, March 1989), (CIA ref. no.: CIA-RDP96-00787R000300170001-2).

Christopher's ARV experiment was published as Smith, C.C., Laham, D. & Moddel, G. (2014) "Stock Market Prediction Using Associative Remote Viewing by Inexperienced Remote Viewers," *Journal of Scientific Exploration*, (Vol.2, No.1, 2014), 7-16.

Chapter 13 - Critical Thinking for Skeptics and Believers Alike

Two good books directly addressing skeptics' attacks are Robert McCluhan's *Randi's Prize: What Sceptics Say about the Paranormal, Why They Are Wrong, and Why It Matters,* (Troubador Publishing Ltd, 2010).; and Chris Carter's *Science and Psychic Phenomena: The Fall of the House of Skeptics,* (Inner Traditions, 2012). Dean Radin's book *The Conscious Universe: The Scientific Truth of Psychic Phenomena,* (Harper Collins, 1997), has a good section on skeptical arguments as well. In addition, an excellent online resource with a wealth of material about the skeptical debate over ESP is the "Skeptical About Skeptics" site (http://www.skepticalaboutskeptics.org/).

Chapter 14 - Remote Perception

The presentiment experiment protocol was first published in Dean Radin's seminal paper, "Unconscious perception of future emotions: An experiment in presentiment." *Journal of Scientific Exploration,* (Vol.11, No.2), 163-180. It is available free online on the Journal of Scientific Exploration website.

A recent summary and meta-analysis of many collected presentiment research studies is Julia Mossbridge, Patrizio Tressoldi, & Jessica Utts, "Predictive Physiological Anticipation Preceding Seemingly Unpredictable Stimuli: A Meta-analysis," *Frontiers in Psychology,* (Vol. 3, Article 390, Oct 2012), 1-18. (Online at http://journal.frontiersin.org/article/10.3389/fnhum.2014.00146/full)

Bryan Williams gives a good overview of the Ganzfeld experiment and the debate surrounding it in his "Revisiting the Ganzfeld ESP Debate: A Basic Review and Assessment," (*Journal of Scientific Exploration,* (Vol. 25, No. 4, 2011), 639-661) (Online at http://www.deanradin.com/FOC2014/Williams2011Ganz.pdf)

A helpful discussion of Jahn and Dunne's precognitive remote perception experiments is available in their book *Margins of Reality,* as well as their article "Information and uncertainty: 25 years of remote perception research" (both cited in the notes for Chapter 12).

Chapter 15 - How Do I Learn Remote Viewing?

See Chapters 11-18 of *Reading the Enemy's Mind* for a detailed, first-person account of what it was like to do training with Ingo Swann.

Chapter 16 - Where do You Go From Here (and how do you get there)?

My view on the state of remote viewing today can be found in "Remote Viewing: State of the Field," *EdgeScience,* No.8, (July-September 2011), 14-16 (Online at http://www.scientificexploration.org/edgescience/edgescience_08.pdf).

INDEX